AF840816

Contents

1 Chapter 1: Luminescent Life

The soft, ethereal glow emanating from the forest floor isn't a trick of the moonlight. It's the product of a fascinating biological phenomenon: fungal bioluminescence. These living lights, often overlooked in the shadowed groves they inhabit, represent a captivating intersection of art and science. The chemical magic within these fungi converts chemical energy into light, a process known as bioluminescence. Luciferins, the light-emitting molecules, react with oxygen in the presence of luciferase enzymes, creating a cold light that ranges from subtle greens to vibrant yellows. This light, a beacon in the darkness, attracts insects that aid in spore dispersal, a crucial part of the fungi's life cycle. Understanding this fundamental process opens a window into the intricate world of these luminous organisms.

These shadowed groves, the preferred habitats of bioluminescent fungi, are ecosystems of subtle beauty and delicate balance. Thick canopies filter sunlight, creating a dim, humid environment perfect for these light-loving organisms. Decaying wood, rich in nutrients, provides a substrate for fungal growth. The interplay of light and shadow, moisture and decay, fosters a unique biodiversity, where bioluminescent fungi play a crucial role. They contribute to nutrient cycling, breaking down organic matter and releasing vital elements back into the ecosystem. The delicate balance of these groves makes them essential havens for bioluminescent fungi. Exploring the nuances of this ecosystem reveals the intricate connections between the fungi and their environment.

Imagine a medium that could capture and interact with this living light, a substance I call "stardust." Conceived as a blend of light-sensitive nanoparticles suspended in a clear, flexible matrix, stardust acts as both a canvas and a conduit. It absorbs the light emitted by the fungi, storing the energy within its nanoparticles. This stored energy can then be re-emitted, allowing for the creation of dynamic, glowing artworks that mimic the rhythmic pulses of the fungi. Furthermore, stardust's flexibility allows it to conform to the irregular surfaces of the fungi and the surrounding environment, integrating seamlessly into the shadowed grove. The potential of stardust as an artistic medium lies in its ability to bridge the gap between the ethereal beauty of bioluminescent fungi and human artistic expression.

Stardust's unique properties go beyond simple light capture. The nanoparticles within the matrix are designed to react to subtle changes in the environment, such as temperature and humidity. This responsiveness allows for the creation of interactive installations, where the glowing artwork shifts and changes in response to the surrounding conditions, reflecting the dynamic nature of the shadowed grove ecosystem. The intensity of the re-emitted light can be modulated by adjusting the concentration of nanoparticles, offering a wide range of artistic possibilities. From subtle, ambient glows to vibrant, pulsating displays, stardust provides a versatile platform for capturing the elusive beauty of fungal bioluminescence.

Beyond its artistic applications, stardust holds potential for scientific research. The nanoparticles can be modified to detect specific chemical compounds emitted by the fungi, providing insights into their metabolic processes. This could lead to a deeper understanding of the biological mechanisms behind bioluminescence and its ecological role. By measuring the intensity and frequency of light emissions captured by stardust, researchers could gain valuable data on the fungi's responses to environmental changes. This data could be used to monitor the health of the shadowed grove ecosystem and inform conservation efforts.

The development of stardust represents a novel approach to interacting with and understanding bioluminescent fungi. It merges the realms of art and science, offering a unique perspective on these elusive organisms. Imagine walking through a shadowed grove, surrounded by artworks that pulsate and glow in harmony with the living fungi. This immersive experience could foster a deeper appreciation for the delicate balance of these ecosystems and inspire new avenues for both artistic expression and scientific inquiry. Stardust, a medium born from the intersection of art and science, promises to unlock the secrets of these luminous wonders and share their ethereal beauty with the world. Its ability to capture, store, and re-emit light offers a powerful tool for both artistic expression and scientific exploration, opening up a new dimension in our understanding of the natural world. The future of stardust, like the fungi it seeks to capture, holds the promise of illuminating the hidden harmonies of nature.

1.1 Fungal Bioluminescence

The ethereal glow emanating from certain fungi isn't magic, but a fascinating biochemical process. Bioluminescence, the production and emission of light by living organisms, is a phenomenon observed across various life forms, from fireflies to deep-sea creatures. In fungi, this captivating display is a result of a specific chemical reaction involving luciferin, a light-emitting molecule, and luciferase, the enzyme that catalyzes the reaction. Oxygen plays a crucial role, interacting with luciferin in the presence of luciferase to produce oxyluciferin, the excited-state molecule responsible for the characteristic glow. The energy released during this transition from the excited state back to the ground state is emitted as photons, the elementary particles of light. This intricate dance of molecules gives rise to the enchanting luminescence that characterizes these remarkable fungi.

The ecological significance of fungal bioluminescence remains an area of ongoing research, but several hypotheses have emerged. One prominent theory posits that the light attracts insects and other arthropods. These drawn-in creatures inadvertently aid in spore dispersal, facilitating the fungi's reproduction. Imagine a tiny mushroom, glowing softly in the darkness, attracting a curious moth. The moth lands, brushing against the spore-bearing surface, and unknowingly carries the fungal spores to new locations. This ingenious strategy utilizes light as a beacon, ensuring the continuation of the fungal species. Another intriguing hypothesis suggests that the light may act as a warning signal, deterring potential predators from consuming the fungi. The glow could signal toxicity or unpalatability, discouraging animals from taking a bite.

The diversity of bioluminescent fungi extends beyond their ecological roles, encompassing a stunning array of colors and intensities. While the iconic "foxfire" often exhibits a greenish-yellow glow, other species emit light ranging from blue to orange. The intensity of the luminescence can also vary, with some species producing a faint, almost imperceptible glow, while others shine brightly enough to illuminate their surroundings. This variation is influenced by several factors, including the species of fungus, its stage of development, and environmental conditions such as temperature and humidity. Understanding these nuances requires careful observation and meticulous documentation, often involving specialized equipment to measure light intensity and spectral emissions.

The biochemical pathways responsible for bioluminescence in fungi are incredibly complex, involving a cascade of enzymatic reactions and intricate molecular interactions. While the basic principles involve luciferin, luciferase, and oxygen, the specific molecules and pathways can differ between fungal species. Researchers are actively working to decipher these intricate mechanisms, employing advanced techniques like gene sequencing and protein analysis to unravel the secrets of fungal light production. Unraveling these complexities not only deepens our understanding of the natural world but also holds potential for diverse applications, from developing new bioluminescent markers for medical research to exploring sustainable lighting solutions inspired by nature's ingenuity.

The study of fungal bioluminescence offers a unique window into the intricate interplay between chemistry, biology, and ecology. By exploring the mechanisms behind this fascinating phenomenon, we gain insights into the evolutionary adaptations of these remarkable organisms. The ability to produce light has evolved independently in various lineages of fungi, suggesting a strong selective advantage in the environments where they thrive. Further research promises to unveil even more secrets about the ecological significance of fungal bioluminescence and its role in shaping the delicate balance of ecosystems.

Beyond the scientific intrigue, the ethereal beauty of bioluminescent fungi captivates artists and nature enthusiasts alike. The soft, otherworldly glow emanating from these organisms evokes a sense of wonder and mystery, inspiring artistic interpretations across various mediums. Photographers venture into darkened forests, equipped with specialized cameras to capture the fleeting luminescence. Painters attempt to recreate the subtle gradations of color and light on canvas, translating the ephemeral beauty into a tangible form. Sculptors use luminescent materials to mimic the fungal glow, creating installations that evoke the enchantment of a shadowed grove.

The fragility of these bioluminescent ecosystems underscores the importance of conservation efforts. Habitat loss due to deforestation and human encroachment poses a significant threat to these unique organisms. Protecting the shadowed groves where these fungi thrive is crucial to preserving their delicate balance and ensuring the survival of these luminous wonders. Education and awareness play a vital role in fostering appreciation for these often-overlooked organisms and promoting responsible stewardship of their habitats. By understanding the intricate connections between fungi, their environment, and the other organisms that share their habitat, we can work towards a future where these enchanting displays continue to illuminate the darkness.

The study of fungal bioluminescence is not just a scientific pursuit; it is a testament to the beauty and complexity of the natural world and a call to protect its wonders for generations to come.

1.2 Shadowed Groves Ecology

The shadowed grove is a realm of hushed stillness, a sanctuary for life that thrives in the absence of direct sunlight. A delicate balance of factors creates this unique ecosystem, a tapestry woven from shade, humidity, and decaying organic matter. The dense canopy overhead filters sunlight, creating a mosaic of dappled light and deep shadows on the forest floor. This subdued illumination plays a crucial role in shaping the microclimate, maintaining consistently high humidity levels vital for fungal growth. The air hangs heavy with the scent of damp earth and decaying leaves, the very essence of decomposition that fuels the life cycle of the bioluminescent fungi we seek. This constant cycle of decay provides the rich substrate these organisms need to flourish, converting the fallen giants of the forest into the building blocks of new life, a hidden alchemy of the grove. Fallen logs, covered in moss and lichen, become nurseries for fungal colonies, their surfaces glowing with an ethereal light. This delicate ecosystem, far from being static, is in a state of perpetual flux, each element intricately connected to the next, a symphony of life played out in the quietest of tones.

The very structure of a shadowed grove fosters specific plant communities adapted to low-light conditions. Shade-tolerant trees, with their broad leaves designed to capture the slightest glimmer of sunlight, dominate the canopy. Beneath them, a diverse understory of shrubs, ferns, and mosses thrives in the filtered light. These plants, in turn, contribute to the humidity of the environment through transpiration, releasing water vapor into the air. Their intertwined root systems bind the soil, preventing erosion and creating a stable foundation for fungal growth. This intricate interplay of flora creates the ideal habitat for bioluminescent fungi, offering them both the sustenance and the environmental conditions they need to thrive. These plants also play a vital role in the distribution of nutrients throughout the grove, their fallen leaves and decaying wood providing a constant source of organic matter for the fungi to decompose.

The soil within a shadowed grove is a complex and dynamic ecosystem in itself. Rich in organic matter from decaying leaves and wood, it provides a fertile breeding ground for a vast array of microorganisms. Bacteria, fungi, and invertebrates work in concert to break down organic material, releasing nutrients back into the soil. This constant cycle of decomposition is essential for the health of the grove, ensuring a steady supply of nutrients for plant growth. The intricate network of fungal hyphae, the thread-like structures that make up the fungal body, permeates the soil, binding it together and improving its structure. These hyphae also play a crucial role in nutrient cycling, transporting essential minerals and water throughout the grove. The very ground beneath our feet teems with unseen life, a hidden world crucial to the balance of the shadowed grove ecosystem.

The interplay of light and shadow is paramount in shaping the ecology of these unique environments. The limited penetration of sunlight creates a gradient of light intensity, influencing the distribution and behavior of organisms within the grove. Bioluminescent fungi, often found in the deepest shadows, utilize their light for a variety of purposes, from attracting insects for spore dispersal to potentially deterring predators. The dappled light patterns on the forest floor create distinct microhabitats, each supporting a unique assemblage of species. This subtle interplay of light and darkness creates a dynamic environment where organisms have adapted to specific niches, maximizing their chances of survival. The interplay of light and dark, life and decay, creates a unique ecosystem within these shadowy realms, a hidden world waiting to be discovered.

The temperature within a shadowed grove remains relatively constant, buffered from the extremes of heat and cold experienced in more exposed areas. The dense canopy insulates the forest floor, creating a cool and stable microclimate. This consistent temperature regime is crucial for the growth and survival of many species, including bioluminescent fungi. Their delicate enzymatic processes, responsible for producing light, are sensitive to temperature fluctuations. The stable environment of the shadowed grove provides the ideal conditions for these complex biochemical reactions to occur. This consistent temperature, coupled with high humidity, creates an environment where these organisms can thrive.

Air movement within the grove is also restricted by the dense canopy. This creates a still, almost stagnant atmosphere, further contributing to the high humidity levels. The reduced airflow also limits the dispersal of spores and other airborne particles, influencing the distribution of fungal species within the grove. This enclosed environment fosters a sense of intimacy and stillness, a world apart from the bustling activity of the outside world. The stillness amplifies the sounds of the forest – the rustling of leaves, the dripping of water, the faint hum of insects. It allows one to fully appreciate the subtle beauty of the shadowed grove, a place where life unfolds at a slower, more deliberate pace. This reduced air movement also plays a significant role in maintaining the stable microclimate, protecting the delicate fungal ecosystems within the shadowed grove.

1.3 The Stardust Medium

Imagine stardust, not as the cosmic dust of distant nebulae, but as a finely milled, light-receptive material, a medium uniquely attuned to capturing and preserving the ephemeral luminescence of fungi. This envisioned medium, meticulously crafted from rare earth elements and photosensitive compounds, possesses the remarkable ability to absorb and retain the bioluminescent emissions of these extraordinary organisms. Think of it as a living canvas, pulsating with the captured light of the fungi, a dynamic artwork evolving in time. This stardust medium is not merely a passive receptor; its composition allows for the manipulation of the captured light, allowing artists to sculpt and shape the glow, creating three-dimensional light sculptures within the shadowed groves.

The stardust medium's unique properties offer a new dimension to artistic expression. Its granular structure, reminiscent of the forest floor itself, can be molded and shaped, mimicking the natural forms of the fungi and their surroundings. By carefully layering the stardust, artists can create intricate reliefs that capture the subtle gradations of light emitted by the fungal colonies. Imagine a sculpted landscape, glowing with the captured light of Mycena chlorophos, its ethereal green luminescence preserved within the stardust matrix. This isn't just capturing an image; it's capturing the essence of the light itself.

The interaction between the stardust medium and the fungal light is a dynamic process. The medium is designed to not just passively absorb the light but to react to it. Its photosensitive components can shift and rearrange in response to the intensity and duration of the fungal emissions, creating slowly evolving patterns and forms within the artwork. This creates a unique temporal element to the art, a dialogue between the living organism and the responsive medium. This responsiveness goes beyond mere recording; it allows for a genuine interaction, a symbiotic relationship between art and nature. Think of the stardust not as a canvas but as a partner in the creative process, a collaborator with the fungi in producing a living, breathing artwork.

The application of the stardust medium extends beyond individual artworks. Consider the potential for large-scale installations within the shadowed groves themselves. By strategically placing stardust-coated surfaces throughout the forest, artists could create immersive environments that respond to the natural rhythms of the fungal bioluminescence. Imagine walking through a grove where the trees themselves seem to glow, their bark covered in stardust, pulsating with the captured light of the fungi at their roots. This creates an ethereal experience, blurring the lines between art, nature, and observer.

This isn't about simply documenting the beauty of bioluminescent fungi; it's about creating a new form of artistic expression that is inextricably linked to the living world. The stardust medium allows artists to capture and manipulate the ephemeral light of these organisms, creating dynamic, evolving artworks that reflect the intricate interplay of light and life within the shadowed groves. Imagine a future where these luminous installations become integral parts of the ecosystem, not just representations of nature but active participants in its ongoing symphony of light and shadow. This is a new frontier for both art and science, a journey into the heart of bioluminescence and its artistic potential.

The challenge lies not just in developing the stardust medium itself but in understanding the nuances of fungal bioluminescence. Each species emits light with different intensities, durations, and spectral qualities. The stardust medium must be adaptable enough to capture this diversity of light, while also being robust enough to withstand the humid environment of the shadowed groves. This requires a deep understanding of both the artistic and scientific principles involved, a merging of disciplines in pursuit of a shared vision. The stardust is not a static material; it's an evolving technology, constantly being refined and adapted to better capture the elusive harmonies of the fungal world.

The potential applications of this medium extend beyond the purely artistic realm. Imagine using stardust-coated sensors to monitor the health and activity of fungal colonies, providing valuable data for scientific research. The subtle changes in light emission could reveal hidden patterns in the fungal life cycle, offering insights into the complex interplay of organisms within the shadowed groves. This merging of art and science creates a feedback loop, where artistic exploration informs scientific inquiry and vice versa. The stardust becomes a bridge between two worlds, fostering a deeper understanding of both the artistic and scientific dimensions of bioluminescent fungi.

Furthermore, the development of the stardust medium presents an opportunity for interdisciplinary collaboration. Artists, scientists, and engineers can work together to refine the properties of the medium, exploring its potential in both artistic and scientific contexts. This collaborative approach fosters innovation and pushes the boundaries of what's possible, blurring the lines between traditional disciplines. Imagine a team of artists and mycologists working together, using the stardust medium to create interactive installations that respond to the real-time bioluminescent activity of fungal colonies. This is a vision of a future where art and science work hand in hand, illuminating the hidden wonders of the natural world.

The ultimate goal of the stardust medium is to create a deeper connection between humans and the natural world. By capturing and manipulating the ephemeral beauty of bioluminescent fungi, we can gain a new appreciation for the intricate interconnectedness of life on Earth. The stardust medium offers a unique opportunity to experience the magic of bioluminescence in a tangible, interactive way, fostering a sense of wonder and respect for the delicate ecosystems that support these luminous organisms. This is not just art; it's a celebration of life, a testament to the power of nature to inspire and amaze. Through the stardust medium, we can glimpse the celestial sphere reflected in the shadowed groves, connecting us to the larger cosmos through the luminescence of the fungal world.

The journey to unlock the full potential of the stardust medium is just beginning. As we continue to explore the properties of this unique material and the intricacies of fungal bioluminescence, we are sure to uncover new and exciting possibilities for artistic expression and scientific discovery. The stardust is not just a medium; it's a gateway, a portal to a world of light and life that is just waiting to be explored. This is an invitation to join us on this journey, to discover the elusive harmonies of the shadowed groves and the celestial sphere they evoke, through the artistic conduit of stardust.

2 Chapter 2: Celestial Connections

Imagine the sprawling network of a mycelium beneath the forest floor, a hidden web of interconnectedness. Consider then, the vast expanse of the cosmos, stars strung together by invisible forces, each influencing the other in a cosmic dance. This chapter explores the concept of the "artistic conduit," the idea that art can bridge these seemingly disparate worlds – the microcosm of the fungal network and the macrocosm of the celestial sphere. We will investigate how the ethereal glow of bioluminescent fungi, captured and manipulated through the medium of stardust, can serve as a visual metaphor for this profound interconnectedness.

Think of the delicate strands of mycelium as analogous to constellations, each point of light a node in a complex network of communication and exchange. Fungi, through their mycelial networks, communicate chemically, sharing resources and information across vast distances beneath the earth. Similarly, celestial bodies exert gravitational influences across unimaginable stretches of space, their positions and movements intertwined in a delicate balance. Stardust, as a hypothetical artistic medium, offers a tangible link between these two realms, allowing us to capture and interpret the luminescence of fungi as a reflection of celestial brilliance.

This artistic conduit isn't simply about visual representation; it's about capturing the essence of connection. It's about understanding the subtle language of light, the way it pulsates and shifts within the fungal network, mirroring the dynamic energy of the cosmos. By utilizing stardust, we can potentially translate these subtle signals, transforming them into tangible artistic expressions that resonate with the viewer on a deeper level. Imagine an art piece that not only depicts the beauty of bioluminescent fungi but also conveys the intricate web of relationships within their hidden world, mirroring the vast interconnectedness of the universe.

Now, let's delve into the specific properties of stardust and how it can facilitate this artistic translation. Imagine a substance capable of absorbing and refracting the light emitted by bioluminescent fungi, preserving its ethereal quality and allowing for manipulation and shaping. Perhaps stardust consists of nanoparticles that react to the unique wavelengths of fungal bioluminescence, amplifying and transforming the light into new forms. The artist, then, becomes a conductor, orchestrating the flow of light, creating patterns and structures that reflect the intricate relationships within the mycelial network and beyond.

Consider the potential of stardust to create three-dimensional representations of mycelial networks, capturing the subtle gradations of light and color that signify different stages of growth and communication. Imagine sculptures that pulsate with the rhythmic emissions of the fungi, mimicking the dynamic flow of energy within the network. These artistic creations would not merely depict the fungi; they would embody their essence, becoming microcosms of the larger interconnected web of life. This connection extends beyond the terrestrial, resonating with the celestial sphere and its own intricate dance of light and energy.

Further, imagine using stardust to create interactive installations where viewers can engage with the captured light of bioluminescent fungi. Perhaps the intensity of the light could respond to touch or sound, creating a feedback loop that reflects the interconnectedness of all things. Such an installation could become a portal, blurring the lines between the observer and the observed, the terrestrial and the celestial, inviting contemplation on the intricate web of life that connects us all.

The artistic conduit, facilitated by stardust, offers a powerful means of exploring the profound parallels between the hidden world of fungi and the vast expanse of the cosmos. It allows us to translate the ethereal language of light into tangible artistic expressions, fostering a deeper understanding of the interconnectedness that permeates all levels of existence. By capturing the subtle brilliance of bioluminescent fungi, we can glimpse the echoes of celestial harmonies, bridging the gap between the microcosm and the macrocosm, the earthly and the divine.

2.1 The Artistic Conduit

Imagine the bioluminescent fungi not merely as organisms, but as living constellations within the shadowed groves. They are points of light, each a tiny beacon pulsating with an inner luminescence, woven together by the unseen threads of their mycelial networks. This interconnectedness, this subtle symphony of light, offers a unique opportunity for artistic exploration – a chance to translate the language of the forest into a visual dialogue that bridges the earthly and the celestial. We aim to use stardust, not as a mere pigment, but as a conduit, a translator of this silent conversation.

Stardust, in this context, can be conceived as a meta-material: a light-sensitive emulsion capable of capturing and preserving the subtle photonic emissions of the fungi. Think of it as a photographic film, but instead of capturing reflected light, it absorbs and retains the bioluminescent glow, allowing for its subsequent manipulation and artistic interpretation. This could involve embedding the stardust within a transparent resin, creating three-dimensional sculptures that hold the light of the fungi within their form. Alternatively, the charged stardust could be applied to canvases, allowing artists to "paint" with the light itself, crafting images that shimmer and shift with the captured bioluminescence.

Consider the possibilities: a swirling nebula of fungal light captured within a crystal sphere, mirroring the celestial bodies above. Or a tapestry woven with stardust threads, depicting the intricate network of a mycelial colony, pulsating with the rhythm of the forest floor. These are not mere representations of the fungi, but embodiments of their luminous essence – a tangible connection to the hidden magic of the shadowed groves. The artist, in this context, becomes a translator, a conduit between the silent language of the fungi and the human eye. This artistic process is not merely about aesthetics; it's about fostering a deeper understanding and appreciation for the delicate ecosystems that harbor these luminous organisms. By capturing the ephemeral beauty of the fungi, we bring their world into focus, highlighting their importance within the larger ecological tapestry. This, in turn, can inspire conservation efforts, ensuring that these glowing wonders continue to illuminate the shadowed groves for generations to come. The artwork becomes a testament to the beauty and fragility of the natural world, a call to action for its preservation.

The artistic conduit, therefore, is a two-way street. It allows us to translate the language of the fungi into a visual form, making it accessible to a wider audience. Simultaneously, it allows us to project our own interpretations and emotions onto the canvas, creating a dialogue between the human and the natural world. This reciprocal exchange is where the true magic lies, where the boundaries between art, science, and the natural world begin to blur. We become active participants in the ongoing conversation of the forest, contributing our own voices to the symphony of light.

Imagine further: immersive installations where viewers can walk through a simulated shadowed grove, surrounded by the captured bioluminescence of the fungi. The stardust could be embedded within interactive surfaces, responding to touch and movement, creating a dynamic and ever-changing luminous environment. This would not only be an artistic experience, but an educational one, offering a glimpse into the hidden world of these fascinating organisms. The viewer becomes immersed in the ecosystem, experiencing the magic of the shadowed groves firsthand.

The potential applications of stardust are vast and varied, limited only by the imagination of the artist. From wearable art that glows with the captured light of the fungi to architectural installations that transform buildings into living canvases, the possibilities are truly endless. The key is to approach the medium with a sense of wonder and respect, recognizing the inherent beauty and fragility of the organisms we are working with. The artistic process becomes a form of reverence, a celebration of the luminous life that thrives within the shadowed groves.

Ultimately, the goal is to create a bridge between the scientific and artistic realms, fostering a deeper appreciation for the natural world through the lens of artistic expression. By harnessing the power of stardust, we can illuminate the hidden wonders of the shadowed groves, sharing their magic with the world and inspiring future generations to protect these precious ecosystems. The artistic conduit becomes a pathway to understanding, a portal to the celestial sphere reflected in the bioluminescent glow of the fungi. It's a journey into the heart of the forest, a testament to the interconnectedness of all living things.

2.2 Mycelial Networks

Beneath the shadowed canopy, a hidden world thrives. A vast, intricate network pulsates with life, connecting individual fungal organisms into a complex, interconnected community. This is the realm of the mycelium, the vegetative part of a fungus, consisting of a network of fine white filaments called hyphae. These thread-like structures weave through the soil and decaying wood, creating a living web that underpins the forest ecosystem. Mycelia are not simply passive absorbers of nutrients; they are active participants in the environment, facilitating communication and resource sharing between trees and other plants. They act as conduits, transporting water, nutrients, and even chemical signals between different species, creating a symbiotic relationship that sustains the entire forest community. Imagine this network as a subterranean internet, constantly exchanging information and resources, maintaining the delicate balance of the shadowed grove.

Consider the implications of this interconnectedness for bioluminescent fungi. Their glow, far from being a solitary phenomenon, becomes a reflection of the entire mycelial network. The light produced by a single mushroom is not isolated but rather a point of expression within a larger, luminous tapestry. The intensity and rhythm of the light may vary depending on the health and activity of the network, revealing the subtle fluctuations and dynamic processes occurring beneath the surface. Changes in environmental conditions, such as moisture levels or nutrient availability, could be reflected in the bioluminescent display, offering a visual representation of the mycelial network's response. This network serves as both the power source and the communication medium for the mesmerizing light show we witness in the shadowed groves.

Now, let us delve deeper into the structure and function of these mycelial networks. The individual hyphae, microscopic in size, collectively form a vast, interconnected web that can extend for miles beneath the forest floor. These hyphae are not simply static structures but are constantly growing, branching, and fusing, creating a dynamic and ever-changing network. This intricate architecture allows for efficient nutrient absorption and transport throughout the fungal colony. Furthermore, the mycelium plays a critical role in decomposition, breaking down organic matter and releasing essential nutrients back into the ecosystem. This process is crucial for maintaining soil fertility and supporting the growth of other organisms. Imagine the hyphae as tiny miners, diligently excavating the soil, extracting valuable resources, and distributing them throughout the interconnected network.

This interconnectedness extends beyond the fungal kingdom. Mycelial networks form symbiotic relationships with the roots of trees and other plants, creating what are known as mycorrhizae. These mutually beneficial associations allow the fungi to access sugars produced by the plants through photosynthesis, while the plants gain access to water and nutrients absorbed by the extensive mycelial network. This intricate web of interactions highlights the interdependence of organisms within the shadowed grove. The health and vitality of the forest are inextricably linked to the functioning of these hidden mycelial networks. They are the foundation upon which the entire ecosystem rests.

The artistic medium of stardust, introduced in the previous chapter, offers a unique opportunity to visualize and interact with these hidden networks. By capturing the subtle variations in light intensity and patterns, stardust can reveal the intricate structure and dynamic processes of the mycelium. Imagine being able to trace the pathways of nutrients and chemical signals as they flow through the network, witnessing the constant communication and exchange between different organisms. Stardust could allow us to visualize the dynamic responses of the mycelium to environmental changes, offering insights into the complex interplay between fungi and their environment. It can also provide a tangible and accessible way for us to connect with the hidden world beneath our feet, fostering a deeper appreciation for the interconnectedness of life in the shadowed groves.

Furthermore, stardust can serve as a powerful tool for scientific inquiry. By precisely measuring and mapping the bioluminescent emissions of fungal networks, we can gain a deeper understanding of their physiological processes and ecological roles. Stardust can provide valuable data on the factors influencing fungal growth, communication, and resource allocation within the mycelial network. This information can help us to develop more effective strategies for conserving these fascinating organisms and the ecosystems they inhabit. By combining artistic expression with scientific investigation, stardust opens up new avenues for exploring the hidden wonders of the fungal world. Finally, consider the artistic possibilities offered by stardust. By capturing the ethereal glow of the mycelium, we can create stunning visual representations of these hidden networks. Imagine luminous sculptures that mimic the intricate architecture of the hyphae, or interactive installations that respond to the dynamic fluctuations of the mycelial network. Stardust can be used to create immersive experiences that transport viewers into the shadowed groves, allowing them to witness the mesmerizing light show of bioluminescent fungi in all its glory. It provides a unique medium for expressing the beauty and complexity of the fungal world, bridging the gap between art and science, and inspiring a deeper appreciation for the interconnectedness of life. This medium allows us to explore the hidden harmonies of the shadowed grove, connecting us to the celestial sphere through the artistry of fungal bioluminescence.

3 Chapter 3: Harmonies of Light

The bioluminescent emissions of fungi are not monochromatic bursts of light; they are complex, nuanced expressions of spectral diversity. Consider the subtle variations in color, ranging from the cool, ethereal greens of O. olearius to the warmer, almost golden hues of some Mycena species. These subtle shifts in wavelength, often imperceptible to the naked eye, reveal a hidden world of complexity. Sophisticated spectrophotometry reveals distinct spectral peaks and troughs, each a fingerprint of the unique chemical reactions driving the bioluminescence. This careful analysis unlocks valuable insights into the luciferin-luciferase systems at play within each fungal species, revealing the interplay of enzymes and substrates that orchestrate the light show. Moreover, examining the spectral emissions in their surrounding environment, considering the filtering effects of leaf litter and canopy cover, unveils the interplay of light and shadow within the shadowed grove, enriching the aesthetic appreciation of these fungal displays.

Beyond their static spectral qualities, these fungal emissions also possess a unique temporal dimension. Some species exhibit a constant, unwavering glow, a steady beacon in the darkness. Others pulse rhythmically, their light waxing and waning in hypnotic cycles. These temporal variations, often synchronized within a colony or even across a larger area, hint at a deeper level of communication and coordination. Time-lapse photography captures the elusive rhythms, revealing patterns that might otherwise go unnoticed. These temporal dynamics, overlaid upon the spectral diversity, add another layer of complexity to the luminous harmonies within the grove, suggesting a sophisticated interplay of biological processes and environmental cues. Imagine the slow pulse of a Panellus stipticus colony against the steady glow of a neighboring Armillaria species, a silent dialogue of light playing out in the undergrowth.

These individual expressions of light, both spectral and temporal, combine to create intricate natural symphonies within the shadowed groves. The interplay of different species, each with its own unique luminescent signature, creates a rich tapestry of light and color. These living light installations, embedded within the complex architecture of the grove, offer a multisensory experience far beyond the visual. The ambient humidity, the scent of decaying leaves, the rustling of unseen creatures – all contribute to the overall symphony. It's a holistic experience, a complete immersion in the luminous world of the fungi. Picture standing within such a grove, enveloped by the soft glow of countless fungal bodies, the air alive with their silent music. The ground, littered with decaying leaves, becomes a canvas for a living masterpiece.

These luminous fungi, acting in concert, can be viewed as complex fungal orchestras. Each individual fruiting body, with its specific spectral and temporal emissions, contributes a unique voice to the overall composition. The conductor, in this biological orchestra, is the complex interplay of environmental factors – temperature, humidity, moonlight, and the presence of other organisms. These factors influence the intensity and rhythm of the bioluminescence, shaping the overall performance. Studying the behavior of these fungal orchestras can provide insights into the complex ecological relationships within the grove. The rise and fall of the luminescent chorus can indicate changes in the microenvironment, offering a glimpse into the hidden life of the forest floor.

Finally, consider the elusive, ephemeral nature of these bioluminescent displays. They are whispers of the grove, fleeting moments of luminous beauty hidden within the darkness. Their elusiveness adds to their allure, making each encounter a precious, almost sacred experience. The challenge of finding and observing these fungi encourages a deep connection with the natural world. It requires patience, attentiveness, and a willingness to embrace the unknown. The reward is a glimpse into a hidden world, a moment of communion with the luminous harmonies of the grove, a connection to the celestial sphere mirrored in the soft glow of these remarkable organisms. And perhaps, in these fleeting encounters, we can glean a deeper understanding of the interconnectedness of all life, from the smallest fungal spore to the vast expanse of the cosmos.

3.1 Spectral Emissions

The bioluminescent emissions of fungi aren't merely a uniform green glow. They represent a complex interplay of chemical reactions, resulting in a surprising spectrum of light. Consider the ghostly blue radiance of Omphalotus nidiformis, also known as the ghost fungus, a stark contrast to the yellowish-green of the more common Panellus stipticus. These subtle variations are often overlooked by the casual observer, but they are crucial to understanding the intricate world of fungal bioluminescence. These spectral differences aren't just aesthetic curiosities; they hold clues to the evolutionary history and ecological roles of these fascinating organisms. Different wavelengths may attract different insects, which play a vital role in spore dispersal. This highlights the importance of observing and documenting these variations with precision, a task that blends scientific inquiry with artistic appreciation.

We must remember that the perceived color of bioluminescence can also be affected by environmental factors. The humidity of the shadowed groves, the temperature of the air, and even the composition of the surrounding substrate can subtly shift the emitted wavelengths. Imagine the emerald glow of a cluster of Mycena chlorophos intensified by the dampness of a recent rain, or the golden hue of a lone Panellus stipticus deepened by the decaying wood it inhabits. These nuances add another layer of complexity to the already fascinating world of fungal light, emphasizing the interconnectedness of these organisms with their environment. Understanding these environmental influences is paramount for accurately capturing and interpreting the spectral emissions of bioluminescent fungi.

Further complicating the spectral landscape is the phenomenon of intraspecific variation. Within the same species, we can sometimes observe different colors of bioluminescence, a testament to the genetic diversity of these fungi. Perhaps a colony of Panellus stipticus displays a range of yellows, from a pale lemon to a rich ochre. This variation could be due to subtle genetic differences within the colony or perhaps micro-environmental factors influencing gene expression. Such observations underscore the dynamic nature of bioluminescence and the constant interplay between genetics and the environment. Documenting these variations provides valuable insights into the adaptive strategies of these remarkable organisms.

Beyond color, the intensity of bioluminescence also varies significantly. Some species emit a faint, ethereal glow, barely perceptible to the naked eye, while others produce a light bright enough to read by. This variation in intensity isn't random. It reflects different ecological strategies and the specific function of bioluminescence for each species. A faint glow might be sufficient to attract tiny insects in a densely forested environment, while a brighter light could be necessary to attract larger dispersers across greater distances.

The duration of bioluminescence is another fascinating aspect of this phenomenon. Some species glow continuously, while others exhibit pulsating or intermittent light emissions. The foxfire, for example, can maintain its glow for extended periods, creating an otherworldly ambiance in the shadowed groves. Other species may only luminesce briefly, perhaps triggered by a specific environmental cue or as part of a complex reproductive strategy. These temporal variations in light emission add another dimension to the already rich tapestry of fungal bioluminescence, highlighting the intricate interplay between light, life, and environment. These rhythmic pulsations of light, often synchronized within a colony, suggest a level of communication and coordination that we are only beginning to understand.

To fully appreciate the spectral emissions of bioluminescent fungi, one must venture into their natural habitat – the shadowed groves. These humid, dark environments are crucial for the survival and luminescence of these fascinating organisms. The lack of ambient light allows the fungal glow to stand out, maximizing its effectiveness in attracting insects and dispersing spores. Stepping into a shadowed grove at night, surrounded by the ethereal glow of bioluminescent fungi, is a truly transformative experience. It offers a glimpse into a hidden world, a realm of living light that challenges our perceptions of the natural world. The experience can be profoundly moving, connecting us to the intricate web of life that thrives in these unique ecosystems. It is within these darkened realms that we can truly appreciate the subtle variations in color, intensity, and rhythm, uncovering the full spectrum of fungal bioluminescence. The delicate interplay of light and shadow reveals the artistry of nature, inspiring awe and wonder. This first-hand observation is invaluable for both scientific study and artistic inspiration, allowing for a deeper understanding and appreciation of the spectral emissions of bioluminescent fungi. By immersing ourselves in these environments, we can begin to unravel the secrets of these luminous organisms and capture their ethereal beauty in our artistic endeavors. Only through careful observation and meticulous documentation can we hope to unlock the full potential of these spectral emissions as both scientific tools and artistic inspiration.

3.2 Elusive Rhythms

Bioluminescent fungi don't simply glow; they pulse, flicker, and undulate with a captivating rhythm. These luminous patterns, often subtle and easily overlooked, are far from random. They represent a complex interplay of biological processes, environmental cues, and potentially even communication within the fungal colony and its surrounding ecosystem. Understanding these elusive rhythms is key to appreciating the full complexity of fungal bioluminescence and harnessing its artistic potential within the shadowed groves. We must move beyond simply observing the glow and delve into the temporal dynamics of this mesmerizing phenomenon.

One of the most intriguing aspects of these rhythmic displays is their variability. Different species of fungi exhibit distinct patterns of luminescence. Some emit a steady, continuous glow, while others pulse with periods ranging from seconds to minutes. This variation is influenced by factors like temperature, humidity, and the availability of nutrients. A sudden drop in temperature, for example, might trigger a burst of luminescence, while a period of drought could diminish the intensity of the glow altogether. These responses reflect the intimate connection between the fungi and their environment, with their light acting as a barometer of the grove's health and a testament to the delicate balance of life within. By carefully observing these variations, we gain valuable insights into the fungal life cycle and the ecological dynamics of the shadowed grove.

Beyond environmental influences, the rhythms of bioluminescence may also play a role in fungal communication. While still largely unexplored, the possibility of fungi using light to signal to one another or to other organisms is tantalizing. Imagine a network of mycelia beneath the forest floor, pulsing with light, conveying information about nutrient availability, potential threats, or even reproductive readiness. This intricate signaling system, hidden beneath the leaf litter, could be a vital component of the grove's ecological web. Unlocking the secrets of this fungal language could revolutionize our understanding of communication in the natural world.

Furthermore, the rhythmic nature of fungal bioluminescence presents unique opportunities for artistic expression using the stardust medium. Imagine capturing these temporal patterns, not as static images, but as dynamic light sculptures that ebb and flow with the rhythm of the fungi. Stardust, with its ability to interact with and amplify the fungal glow, offers a pathway to translating these elusive rhythms into a tangible artistic form. By manipulating the properties of stardust, we could create installations that respond to changes in the environment, mirroring the dynamic nature of the fungi themselves and transforming the shadowed grove into a living art canvas.

The study of these rhythms also demands meticulous observation and documentation. Long-exposure photography, time-lapse videography, and sensitive light meters become essential tools for capturing the subtle changes in luminescence over time. By carefully analyzing these recordings, we can decipher the underlying patterns and uncover the factors influencing them. This detailed data collection provides a scientific foundation for artistic interpretation, allowing us to move beyond mere representation and create art that is both aesthetically captivating and scientifically informed. The intersection of science and art, in this context, becomes a powerful tool for both understanding and appreciating the hidden wonders of the shadowed grove.

Moreover, exploring the rhythmic aspects of fungal bioluminescence requires an interdisciplinary approach. Collaborations between mycologists, ecologists, artists, and engineers are crucial for unraveling the complex interplay of biological, environmental, and artistic factors. By combining their expertise, these diverse perspectives can illuminate the hidden meanings within the elusive rhythms of the fungal glow. Such collaborations can lead to innovative artistic creations that not only showcase the beauty of bioluminescence but also deepen our understanding of the intricate web of life within the shadowed grove. Imagine an interactive installation that translates the rhythmic pulsations of the fungi into sound, creating a symphony of light and sound that immerses the viewer in the magic of the grove.

Finally, understanding the rhythms of fungal bioluminescence can also inform conservation efforts. By monitoring these patterns, we can gain valuable insights into the health and resilience of fungal populations. Changes in the rhythm of the glow could serve as early warning signs of environmental stress, allowing for timely interventions to protect these delicate ecosystems. The art created with stardust, inspired by these rhythms, can then become a powerful tool for raising awareness and advocating for the preservation of these unique and enchanting environments. The luminous language of the fungi, translated through art, can speak to the importance of protecting these shadowed groves for future generations.

3.3 Natural Symphonies

Imagine standing within a shadowed grove, enveloped by an ethereal luminescence. Observe how the soft glow of bioluminescent fungi punctuates the darkness, each fungal colony pulsing with its own unique rhythm. This rhythmic interplay, a symphony of light orchestrated by nature itself, creates an immersive sensory experience. These subtle variations in light intensity and frequency, often imperceptible to the casual observer, reveal a complex communication network within the fungal community. These organisms, seemingly static, engage in a dynamic exchange of information, expressed through the language of light. The grove becomes a living canvas, a stage for a performance orchestrated by the unseen forces of nature.

These luminous emissions are not merely random flickers; they possess a distinct structure, a pattern that unfolds over time. Think of a conductor leading an orchestra, each musician contributing their unique sound to create a harmonious whole. Similarly, each fungal colony, with its individual rhythm of bioluminescence, contributes to the overall symphony of the grove. This intricate interplay of light and shadow, rhythm and pause, creates a mesmerizing spectacle, a testament to the complex communication networks operating within the fungal community. This network extends beyond individual colonies, connecting the entire ecosystem in a web of light-based interactions. The forest floor becomes a tapestry of interconnectedness, woven with threads of bioluminescent communication.

The spectral emissions of these fungi vary considerably, contributing to the richness and complexity of the natural symphony. Some species emit a cool, ethereal green, while others radiate a warmer, more inviting amber hue. This diversity of color adds another layer to the symphony, creating a visual tapestry that shifts and changes with the rhythms of the fungal orchestra. Imagine the subtle interplay of greens and blues, punctuated by the occasional flash of yellow or orange, each color representing a different species, a different voice in the chorus. This visual symphony is not static; it evolves over time, responding to changes in the environment and the internal dynamics of the fungal community.

Furthermore, the intensity of the light emitted by these fungi is not constant. It fluctuates, creating a dynamic interplay of light and shadow, a visual rhythm that pulses through the grove. This rhythmic fluctuation is not random; it follows a pattern, a tempo that is unique to each species. Some species pulse slowly, their light gradually intensifying and then fading, while others exhibit a more rapid, staccato rhythm. These variations in tempo create a complex polyrhythm, adding depth and intricacy to the overall symphony of the grove. Consider how these rhythms might interact, overlapping and intertwining to create a constantly evolving soundscape of light.

To truly appreciate the natural symphonies of bioluminescent fungi, one must immerse oneself in the environment, becoming attuned to the subtle nuances of light and shadow. Patience is key. Allow your eyes to adjust to the darkness, to the subtle gradations of luminescence that paint the forest floor. Listen with your eyes, attuning yourself to the rhythm of the glowing fungi. Observe how the light changes over time, how it responds to the environment, to the presence of other organisms. In this immersive experience, you will begin to understand the complex language of light, the intricate communication network that connects the fungal community.

This understanding extends beyond the purely aesthetic. By observing these natural symphonies, we gain insight into the ecological roles of bioluminescent fungi. Their light attracts insects, which aid in spore dispersal, contributing to the propagation of the species. The rhythmic pulsing may serve as a form of communication between fungal colonies, coordinating their activities and optimizing resource utilization. The spectral emissions may play a role in attracting specific types of insects, further refining the symbiotic relationships within the ecosystem. Each flash of light, each subtle variation in intensity and color, tells a story about the intricate web of life that connects the organisms of the shadowed grove.

As you continue your exploration of these natural symphonies, consider the broader implications of this phenomenon. Bioluminescent fungi offer a unique window into the world of fungal communication, a world that remains largely unexplored. By studying these organisms, we can gain a deeper understanding of the complex interactions that shape ecosystems, the intricate networks of communication that connect living organisms. This knowledge can inform our conservation efforts, helping us to protect these delicate ecosystems and the fascinating organisms that inhabit them. The symphony of light emanating from the shadowed grove is not merely a spectacle of nature; it is a testament to the interconnectedness of life, a reminder of the intricate web of relationships that sustains our planet. By listening to these natural symphonies, we can deepen our appreciation for the wonders of the natural world and our understanding of the complex processes that shape life on Earth.

3.4 Fungal Orchestras

Imagine stepping into a shadowed grove, the air thick with humidity and the scent of damp earth. As your eyes adjust to the dimness, points of light begin to emerge, like scattered stars fallen to earth. These are not celestial bodies, but the fruiting bodies of bioluminescent fungi, each a miniature beacon pulsing with an ethereal glow. Consider them not as individual organisms, but as instruments in a vast, silent orchestra, each contributing its unique luminescence to a symphony of light. This is the concept of the fungal orchestra, a harmonious interplay of bioluminescent organisms creating a breathtaking natural spectacle.
The composition of this living orchestra is dictated by the distribution of different fungal species within the grove. Some clusters might be dominated by a single species, creating a uniform field of light, perhaps a soft green or a vibrant blue. Other areas may host a diverse array of species, each emitting a different hue, resulting in a mosaic of color. The intensity of the light can vary as well, with some fungi pulsing rhythmically, while others maintain a steady glow. These

variations in color, intensity, and rhythm create a dynamic and ever-changing visual experience, like a musical score played out in light. This interplay is not random, but influenced by environmental factors like temperature, humidity, and the presence of other organisms.

The substrate upon which the fungi grow plays a crucial role in the overall composition. A fallen log might become a stage for a miniature light show, with different species colonizing various parts of the decaying wood. The texture and moisture content of the log influence which species thrive, thus shaping the visual display. Similarly, the forest floor itself can become a canvas, with clusters of bioluminescent fungi emerging from the leaf litter, their light filtered through the delicate network of roots and decaying leaves. This interplay between the fungi and their environment adds another layer of complexity to the fungal orchestra, highlighting the interconnectedness of life within the shadowed grove.

These fungal orchestras are not merely static displays of light. They are dynamic systems, constantly changing in response to their surroundings. A sudden drop in temperature might dim the glow of some species, while a passing animal might trigger a wave of bioluminescence as the fungi are disturbed. Even the slightest breeze can create a shimmering effect, as the light from the fungi dances through the leaves. This dynamic nature makes each visit to a fungal orchestra a unique experience, a fleeting glimpse into the hidden world of bioluminescent life. Furthermore, the ephemeral nature of these displays emphasizes the importance of preservation efforts, ensuring that future generations can witness these natural wonders.

The concept of the fungal orchestra invites us to shift our perspective, to see the forest not as a collection of individual organisms, but as a complex and interconnected system. Each fungal species, each fallen log, each patch of leaf litter plays a role in the overall symphony of light. By understanding these connections, we can gain a deeper appreciation for the delicate balance of life within the shadowed grove. This holistic view is essential for effective conservation efforts, as it highlights the interconnectedness of all living things within this unique ecosystem. Imagine being able to "conduct" these fungal orchestras, influencing their light displays through carefully manipulated environmental factors. The possibilities for artistic expression are vast, offering a unique blend of science and art.

Observing a fungal orchestra is a multisensory experience. The soft glow of the fungi is often accompanied by the subtle sounds of the forest – the rustling of leaves, the chirping of insects, the distant call of an owl. These sounds, combined with the visual spectacle of the bioluminescent fungi, create an immersive experience that connects us to the natural world in a profound way. This connection fosters a sense of wonder and awe, reminding us of the beauty and complexity of life on Earth. The fungal orchestra becomes a metaphor for the interconnectedness of all living things, a reminder that we are all part of a larger ecosystem.

Beyond the aesthetic beauty, studying these fungal orchestras provides valuable scientific insights. By observing the interactions between different species, scientists can learn more about the ecological roles of bioluminescent fungi. This knowledge can be applied to conservation efforts, helping to protect these delicate ecosystems and the organisms that call them home. Moreover, the study of bioluminescence in fungi could have applications in other fields, such as medicine and biotechnology. The potential for discovery is vast, and the fungal orchestra serves as a powerful reminder of the untapped potential of the natural world. The future of these luminous wonders depends on our ability to appreciate and protect them, ensuring that their ethereal harmonies continue to echo through the shadowed groves for generations to come.

3.5 Whispers of the Grove

Within the hushed confines of the shadowed groves, a symphony of light unfolds. Not a cacophony, but a delicate interplay of bioluminescent pulses, a language whispered between the fungi and the darkness. These luminous emissions are not merely a spectacle; they are a form of communication, a complex interplay of signals still largely undeciphered by human understanding. Imagine standing amidst this ethereal glow, the air thick with the scent of damp earth and decaying leaves, the only sound the gentle rustling of unseen creatures. The fungi, in their silent luminescence, speak of intricate relationships with their environment, with the insects they attract, and perhaps even with each other. This is the true "whisper of the grove" – a subtle conversation carried on light waves, a biological Morse code awaiting our interpretation.

This luminous dialogue takes on diverse forms. Some species emit a steady glow, a constant beacon in the undergrowth. Others pulse rhythmically, their light waxing and waning in patterns that suggest a hidden timekeeping mechanism. Still others respond to touch or disturbance with bursts of light, like startled fireflies in the forest floor. These varied expressions of bioluminescence suggest a complexity far beyond simple metabolic processes. They hint at a nuanced interplay of environmental factors, internal biological clocks, and perhaps even a rudimentary form of sensory perception. Consider the possibility that these light patterns communicate information about the health of the mycelium, the presence of predators, or the availability of nutrients.

To truly appreciate the whispers of the grove, one must attune their senses to the subtle nuances of light and shadow. The human eye, adapted to the brightness of day, must adjust to the dimness of the forest floor, allowing the faintest glimmers to register. This process of adaptation, of slowing down and focusing, is crucial to experiencing the full richness of the fungal light show. It's a form of sensory meditation, a quieting of the mind to allow the whispers to emerge from the background noise. Close your eyes for a moment, then open them slowly, allowing your pupils to dilate and absorb the subtle gradations of luminescence. Notice how the light interacts with the textures of the surrounding environment, casting shadows that dance and shift with every gentle breeze.

The experience of witnessing these bioluminescent displays is deeply personal and evocative. It's a moment of connection with the natural world, a reminder of the hidden wonders that exist just beyond our everyday perception. For the artist, these whispers of the grove become a source of inspiration, a challenge to translate the ephemeral language of light into tangible forms. How can one capture the delicate interplay of luminescence and shadow, the rhythmic pulsing of fungal light, the sense of mystery and wonder that pervades the shadowed groves? This is the artistic challenge presented by the bioluminescent fungi – to create a visual echo of the whispers, a tangible representation of the intangible.

The exploration of these luminous organisms requires not only artistic sensitivity but also a deep respect for the delicate ecosystems they inhabit. The shadowed groves, with their unique microclimates and intricate web of life, are fragile environments easily disrupted by human interference. As we delve into the mysteries of fungal bioluminescence, we must tread lightly, minimizing our impact and prioritizing the preservation of these extraordinary habitats. This means adhering to ethical guidelines for collecting and cultivating specimens, minimizing disturbance to the surrounding environment, and promoting conservation efforts to protect these natural wonders for future generations. The whispers of the grove are a precious gift, and it is our responsibility to ensure that they continue to resonate through the forest for years to come. The ethereal beauty of these luminous fungi serves as a powerful reminder of the interconnectedness of life and the importance of preserving the delicate balance of our planet's ecosystems.

4 Chapter 4: Stardust as a Canvas

Imagine the forest floor as your canvas, a tapestry woven with shadows and the faint, ethereal glow of bioluminescent fungi. Stardust, in this context, isn't the cosmic dust of distant nebulae, but a specially formulated, light-sensitive medium designed to capture and preserve these fleeting luminescences. This chapter unveils the techniques for utilizing stardust, transforming the ephemeral beauty of fungal light into lasting works of art. Think of it as a photographic emulsion, but instead of reacting to the broad spectrum of visible light, stardust is uniquely attuned to the specific wavelengths emitted by these extraordinary fungi. Its fine granularity allows for incredibly detailed captures, preserving the subtle gradations and patterns of the fungal glow. The application process demands a delicate touch. Prepare the stardust emulsion under red light – a wavelength that won't prematurely activate the medium – and carefully apply a thin, even layer to your chosen substrate. This could be a traditional canvas, a sheet of prepared paper, or even a piece of natural wood gathered from the forest floor itself, enhancing the connection between the artwork and its source.

Once prepared, venture into the shadowed groves during the peak luminescence period, typically after nightfall when the fungi's glow is most pronounced. Position your stardust-coated canvas near the fungal colony, ensuring minimal disturbance to the delicate ecosystem. The exposure time will vary depending on the species and the intensity of their light. Some species emit a constant, steady glow, while others pulse rhythmically, offering a different kind of artistic challenge. Experiment with different exposure durations, capturing the subtle nuances of the fungal light. You might be surprised by the range of effects achievable, from faint, ethereal glows to vibrant, almost pulsating patterns. Remember, this is a collaborative process between artist and fungi; allow the natural rhythms of the organism to guide your artistic expression. Patience is paramount in this endeavor. The faint light of these fungi requires extended exposure times, often stretching into hours.

The development process is equally crucial. Once the exposure is complete, carefully transport your stardust-coated canvas back to your darkroom, again under red light. The development process uses a specialized solution that reacts with the stardust, permanently fixing the captured luminescence. The resulting image isn't a photograph in the traditional sense, but rather a luminous imprint of the fungal glow, a tangible record of its ephemeral beauty. This process allows you to capture not just the visual form of the fungi, but also the essence of their light, their ethereal glow preserved within the stardust matrix. The final piece becomes a luminous echo of the forest, a testament to the hidden wonders of the natural world.

Consider incorporating natural elements into your artwork. Perhaps embed dried leaves or small twigs collected from the grove within the stardust emulsion before exposure. These incorporated elements will create unique textures and patterns, further intertwining the artwork with its environment. Experiment with layering techniques, exposing the stardust to multiple fungal colonies, each with its unique spectral signature. This will create complex, multi-layered images, capturing the interplay of light and shadow within the grove. Don't be afraid to push the boundaries of this medium. The interplay between art and science, between human creativity and natural phenomena, is where true innovation lies. Explore the possibilities of mixed media, combining stardust captures with other artistic techniques, such as painting or drawing. Imagine using luminescent pigments to enhance the fungal glow, creating an even more vibrant and ethereal effect. The possibilities are as vast and varied as the fungal kingdom itself.

Stardust offers a unique opportunity to bridge the gap between scientific observation and artistic expression. It allows us to capture the elusive beauty of bioluminescent fungi, transforming their ephemeral light into a tangible form. More than just a technique, working with stardust becomes an immersive experience, a journey into the hidden world of luminescent life. Embrace the challenges, the unexpected discoveries, and the quiet beauty of these shadowed groves. Let the fungi be your guide, their light your inspiration. Through the medium of stardust, we can not only capture the light of these fascinating organisms, but also share their ethereal beauty with the world. By combining scientific understanding with artistic vision, we can create works that both illuminate and inspire, fostering a deeper appreciation for the hidden wonders of the natural world. Remember, the true magic of this medium lies not just in the capturing of light, but in the stories it allows us to tell.

4.1 Capturing Light

The ethereal glow of bioluminescent fungi presents unique challenges for capture. Unlike conventional photography subjects, these delicate emissions require specialized techniques and equipment. Consider the sensitivity of your camera sensor. A high ISO setting is crucial, allowing you to capture the faintest luminescence without excessive noise. However, be mindful of the balance. Too high an ISO can introduce grain, detracting from the ethereal quality of the light. Experimentation is key, finding the sweet spot between sensitivity and clarity for your specific equipment and environment.

Think beyond traditional long exposures. While long exposures are essential for gathering enough light, they can also blur the intricate details of the fungal structures. Explore bracketing your exposures. Capture a range of exposures, from short to long, allowing you to combine them later in post-processing. This technique preserves both the delicate glow and the fine textures of the fungi. Consider the unique characteristics of stardust as a medium. Its light-reactive properties can enhance the captured luminescence, creating an otherworldly effect. Experiment with different concentrations and application methods of stardust to discover its full potential.

Lens choice is paramount. A fast lens, with a wide aperture (such as f/1.4 or f/2.8), is essential for gathering the maximum amount of available light. The wider the aperture, the more light reaches the sensor, allowing for shorter exposure times and reducing the risk of motion blur. Consider the focal length. A macro lens allows you to capture the intricate details of the fungal structures, while a wider-angle lens can capture the broader context of the shadowed grove environment.

Light pollution is the enemy of bioluminescence photography. Even the faintest ambient light can overpower the delicate glow of the fungi. Seek out the darkest possible locations, far from city lights and other sources of artificial illumination. The new moon phase is ideal, minimizing the ambient light from the sky. Utilize light-blocking materials. A dark cloth draped over your camera and tripod can further minimize stray light, ensuring that your sensor captures only the desired bioluminescence.

Preparing for a bioluminescence photography expedition requires meticulous planning. Research the specific species of fungi you intend to photograph. Understand their preferred habitats, peak luminescence periods, and any environmental sensitivities. Consult with local mycologists or naturalists for guidance on suitable locations and ethical considerations. Pack your equipment carefully. Ensure you have spare batteries, memory cards, and cleaning supplies. The damp environment of shadowed groves can be challenging for electronic equipment, so take precautions to protect your gear.

Patience is a virtue in bioluminescence photography. These elusive organisms often reveal their glow only under specific conditions. Be prepared to wait, observe, and adapt. The rewards are well worth the effort. The ethereal beauty of bioluminescent fungi, captured through the lens and enhanced by stardust, offers a unique and captivating artistic experience. Embrace the challenge and let your creativity be guided by the whispers of the grove.

Remember the importance of composition. While capturing the light is crucial, consider the overall aesthetic of your image. Use natural elements, such as branches or leaves, to frame the fungi and create visual interest. Experiment with different angles and perspectives. Sometimes, shooting from a low angle can emphasize the glow and create a sense of mystery.

Post-processing can further enhance the beauty of your bioluminescent images. Subtle adjustments to brightness, contrast, and saturation can bring out the delicate details of the fungi and the ethereal quality of their light. However, avoid over-processing. Strive to maintain a natural look, preserving the magic of the original scene.

Document your process meticulously. Keep detailed notes on the location, time, weather conditions, and camera settings for each photograph. This information is invaluable for future expeditions and for sharing your experiences with others. Consider the ethical implications of your work. Bioluminescent fungi are delicate organisms and their habitats are often fragile. Minimize your impact on the environment. Avoid disturbing the fungi or their surroundings. Practice Leave No Trace principles, packing out everything you pack in.

Sharing your images can inspire others to appreciate the beauty and wonder of bioluminescent fungi. Consider exhibiting your work in galleries or online platforms. Engage with other artists and scientists interested in bioluminescence. Through collaboration and shared passion, we can further the understanding and appreciation of these remarkable organisms. The journey of capturing bioluminescent light is a blend of scientific observation, artistic expression, and a deep respect for the natural world. Embrace the challenge and allow yourself to be captivated by the elusive harmonies of light.

4.2 Artistic Expression

Consider the delicate interplay of light and shadow, the ethereal glow of bioluminescent fungi against the deep, velvety darkness of the shadowed grove. Stardust, as a medium, allows us to interact with this luminescence, to capture its fleeting beauty and translate it onto a tangible canvas. Imagine a piece where the stardust, imbued with the captured light of Mycena luxaeterna, traces the delicate gills of the mushroom, its glow radiating outwards, mimicking the fungus's natural luminescence. The darkness of the backing material, perhaps a rich, dark fabric woven with luminescent threads, amplifies the fungal glow, creating a three-dimensional effect that draws the viewer into the shadowed grove's depths.

Think beyond mere representation. Stardust enables us to explore the abstract concepts surrounding fungal bioluminescence. The interconnectedness of the mycelial network, for instance, could be depicted using stardust trails weaving across the canvas, intersecting and diverging, reflecting the complex communication systems within the fungal colony. These trails, subtly glowing with different intensities and hues, could represent the flow of nutrients and information through the network, creating a visual representation of the hidden life pulsating beneath the forest floor.

The rhythmic pulsations of certain species, like the Panellus stipticus, offer another avenue for artistic exploration. Stardust, applied in layers of varying density and infused with the specific light signature of the chosen species, can be used to create a visual representation of these rhythmic emissions. The artwork could pulsate gently with light, mimicking the natural rhythm of the fungus, evoking a sense of wonder and awe in the observer. By incorporating sensors and microcontrollers, this pulsation could even be synchronized with recordings of the grove's ambient sounds, further blurring the lines between art and the natural world.

Explore the potential of mixed media. Combine stardust with other materials – perhaps fragments of bark, dried leaves, or even cultivated mycelium – to create textured, multi-layered artworks. Imagine a piece where stardust-infused resin flows over a bed of preserved moss, the embedded glow of the stardust mimicking the scattered light of fungi on the forest floor. Incorporating natural elements into the artwork not only adds another dimension to the visual experience, but also reinforces the connection between the art and the environment it represents.

Consider the canvas itself. Move beyond traditional surfaces and explore the possibilities of unconventional materials. A fallen log, meticulously cleaned and treated, could become a canvas for a stardust depiction of the fungal colonies that might inhabit it. A smooth, dark stone, embedded with stardust mimicking the veins of mycelium, could become a sculptural representation of the hidden fungal network. The choice of canvas can further enhance the narrative of the artwork, adding layers of meaning and deepening the viewer's engagement.

Don't be afraid to experiment with scale. While intricate, small-scale pieces can capture the delicate beauty of individual fungi, large-scale installations can evoke the immersive experience of wandering through a shadowed grove. Imagine a darkened room, the walls lined with fabric imbued with stardust, the glowing patterns mimicking the intricate network of mycelium. Scattered throughout the room, sculptures created from stardust-infused resin could represent clusters of bioluminescent fungi, their ethereal glow illuminating the space, creating a truly immersive and awe-inspiring experience.

Remember, the goal is not simply to replicate the appearance of bioluminescent fungi, but to capture their essence, their ethereal beauty, and the magic they bring to the shadowed groves. Stardust, as a medium, allows us to explore this magic in new and exciting ways, pushing the boundaries of artistic expression and deepening our connection with the natural world. Through careful observation, experimentation, and a willingness to embrace the unexpected, we can unlock the full artistic potential of this remarkable medium and create works that inspire wonder and awe. Embrace the challenge. Let the elusive light of these extraordinary organisms guide your artistic journey. Allow the stardust to become a conduit, translating the whispers of the grove into a language that speaks to the soul. Explore the textures, the colors, the rhythms, and the very essence of bioluminescence, transforming the ephemeral into the tangible, the unseen into the seen. Through your art, illuminate the hidden wonders of the shadowed grove and share its magic with the world.

5 Chapter 5: The Elusive Fungi

The pursuit of bioluminescent fungi is an adventure in itself. These elusive organisms, often tucked away in the deepest recesses of old-growth forests, demand patience, persistence, and a touch of luck to uncover. Begin your search in environments known to favor fungal growth: damp, decaying wood, leaf litter, and areas with high humidity. Focus on regions with a history of bioluminescent fungi sightings, utilizing online resources and mycological societies as valuable tools. Night expeditions are essential, as the ethereal glow of these fungi becomes most apparent in the absence of sunlight. Equip yourself with a red-light headlamp to preserve your night vision without disturbing the fungi's delicate light display. Careful observation is key; scan the forest floor, fallen logs, and even the undersides of leaves. Document potential sightings with photographs, noting the precise location, surrounding environment, and any associated flora or fauna. This meticulous documentation will be invaluable for future research and preservation endeavors.

Cultivating bioluminescent fungi presents a unique set of challenges. Sterile laboratory conditions are paramount to prevent contamination and ensure successful growth. Start with a pure culture obtained from a reputable mycological source, as attempting to cultivate from wild specimens can be complex and often unsuccessful. Agar plates enriched with nutrients specific to the chosen fungal species provide the initial growth medium. Temperature, humidity, and light exposure must be carefully controlled, mimicking the natural environment of the fungi. As the mycelium develops, transfer it to sterilized substrates like wood chips or grain. Patience is crucial during this process; the slow growth rate of bioluminescent fungi necessitates weeks, sometimes months, of attentive care. Regular monitoring for contamination and adjustments to environmental factors are vital for achieving a healthy, glowing culture. It's a rewarding experience to witness the subtle emergence of bioluminescence in a controlled setting, a testament to the intricate balance of nature replicated in the laboratory.

Preservation of bioluminescent fungi habitats is of paramount importance, considering their fragile ecosystems and the increasing threats posed by habitat loss and environmental degradation. Collaborate with local conservation organizations and land management agencies to identify and protect areas where these fungi thrive. Support initiatives aimed at sustainable forestry practices, minimizing disturbance to the delicate balance of shadowed groves. Educate the public about the importance of these unique ecosystems, fostering a sense of stewardship and responsibility. Advocate for the creation of protected areas, safeguarding the long-term survival of bioluminescent fungi and the intricate web of life they support. Citizen science projects can play a crucial role in monitoring the distribution and abundance of these fungi, providing valuable data for conservation efforts. By combining scientific research with community engagement, we can collectively ensure the preservation of these luminous wonders for generations to come.

Ethical considerations are paramount when studying and interacting with bioluminescent fungi. Minimize disturbance to their natural environment during observation and collection. Obtain necessary permits and permissions before entering protected areas or conducting research. When collecting samples, limit the quantity taken and ensure it does not negatively impact the fungal population. Respect the delicate balance of the ecosystem, avoiding any actions that could disrupt the intricate web of life within the shadowed groves. Share research findings and data openly, promoting collaboration and transparency within the scientific community. Engage in responsible tourism practices, educating visitors about the importance of these unique organisms and their habitats. By adhering to ethical guidelines, we can ensure the long-term survival of bioluminescent fungi while fostering a deeper understanding and appreciation for their remarkable properties.

The future of bioluminescent fungi research holds immense promise across various fields. Advancements in genetic sequencing and molecular biology are unraveling the intricate mechanisms behind their bioluminescence, potentially leading to breakthroughs in biomedical imaging and bioengineering. Exploring the symbiotic relationships between these fungi and other organisms could reveal insights into ecosystem dynamics and the interconnectedness of life in shadowed groves. Further investigation into the ecological roles of bioluminescence may uncover new strategies for sustainable agriculture and environmental monitoring. The development of novel bioluminescent materials inspired by these fungi could revolutionize lighting technologies, offering energy-efficient and environmentally friendly alternatives. The intersection of art and science continues to flourish, with artists and scientists collaborating to explore the aesthetic and scientific dimensions of these luminous organisms. By fostering interdisciplinary collaboration and supporting continued research, we can unlock the full potential of bioluminescent fungi, illuminating pathways towards a brighter and more sustainable future.

5.1 Finding the Glow

The pursuit of bioluminescent fungi isn't a casual stroll through sun-drenched meadows. These elusive organisms thrive in the deepest recesses of shadowed groves, often requiring a keen eye, a touch of adventurous spirit, and a healthy dose of patience to uncover. Start your search by researching likely habitats. These fungi often prefer humid, decaying wood in old-growth forests where sunlight barely penetrates. Look for areas with dense leaf litter, fallen logs, and a consistent moisture source, perhaps a nearby stream or spring. Consult mycological guides specific to your region, focusing on species known to exhibit bioluminescence. Online forums and local mycological societies can also provide invaluable insights into recent sightings and specific locations known to harbor these ethereal beauties.

Remember, preparation is key to a successful expedition. Equipping yourself with the right tools will greatly enhance your chances of discovery. A reliable headlamp with a red-light setting is essential. Red light preserves your night vision and minimizes disturbance to nocturnal wildlife, including the fungi themselves. A sturdy walking stick will assist in navigating uneven terrain, while a magnifying glass will allow for close inspection of potential specimens. Durable, waterproof boots are a must, as the forest floor can be damp and muddy. Carry a GPS device and a compass, ensuring you can always find your way back, particularly as darkness descends. Finally, bring a notebook and a camera capable of long exposures to document your findings.

Once in the field, train your senses to the subtle cues these fungi offer. During daylight hours, search for the substrate they prefer. Look for decaying wood, especially logs in advanced stages of decomposition. Note the presence of other fungi, as certain species often coexist with bioluminescent varieties. While the glow itself isn't visible during the day, identifying suitable habitats significantly narrows your search area. As dusk settles, the real hunt begins. Slowly and methodically scan the forest floor, allowing your eyes to adjust to the diminishing light. The bioluminescence, often a soft green or blue hue, may appear as faint flickers or a subtle, ethereal glow emanating from the wood. Move cautiously, avoiding disturbing the delicate ecosystem.

Don't be discouraged if your initial forays prove fruitless. Finding these elusive fungi requires persistence and a touch of luck. Vary your search locations, exploring different groves and microclimates within the forest. Return to promising spots at different times of the year, as bioluminescence can be influenced by temperature, humidity, and lunar cycles. Consider joining organized night hikes or bioluminescence walks led by experienced naturalists. These guided excursions can provide valuable tips and increase your chances of witnessing this magical phenomenon. The thrill of discovering these hidden gems, however, makes the effort worthwhile.

The experience of encountering bioluminescent fungi in their natural habitat is truly unforgettable. It's a moment of connection with the hidden wonders of the natural world, a testament to the beauty that thrives in darkness. Remember to approach these organisms with respect, minimizing your impact on their environment. Avoid touching or disturbing the fungi, and leave the surrounding area undisturbed. The goal is to observe and appreciate their ethereal glow, leaving them to flourish for future generations.

Documenting your discoveries is a vital aspect of the pursuit. Photography can capture the delicate beauty of bioluminescent fungi, but it requires specialized techniques. Use a camera with manual settings, allowing you to control aperture, shutter speed, and ISO. A tripod is essential for long exposures, which are necessary to capture the faint light emitted by the fungi. Experiment with different exposure times and apertures to find the optimal balance between brightness and detail. Focus carefully, as the low light conditions can make achieving sharp images challenging.

Beyond the technical aspects, consider the artistic potential of your photographs. Strive to capture the ethereal quality of the glow, conveying the sense of wonder and mystery that surrounds these organisms. Experiment with composition, using elements of the surrounding environment to frame the fungi and create a visually compelling narrative. Consider incorporating other natural light sources, such as moonlight or starlight, to add depth and dimension to your images.

The pursuit of bioluminescent fungi isn't solely about the destination; it's about the journey itself. It's about immersing yourself in the beauty and mystery of the natural world, attuning your senses to the subtle rhythms of the forest. It's about the thrill of discovery and the quiet contemplation that comes from witnessing the ephemeral magic of bioluminescence. Embrace the challenge, be patient, and prepare to be amazed by the hidden wonders that await you in the shadowed groves. This experience will not only enrich your understanding of the natural world but also inspire your artistic endeavors. The ethereal glow of these fungi awaits, beckoning you to explore the hidden harmonies of the forest.

5.2 Cultivation Trials

Successful cultivation of bioluminescent fungi outside their natural habitat presents a unique set of challenges. These organisms, accustomed to the specific conditions of shadowed groves, often prove difficult to replicate in a laboratory setting. Understanding their nuanced relationship with their environment is crucial. Consider the substrate. In their natural environment, these fungi thrive on decaying wood, leaf litter, and other organic matter specific to the forest floor. Replicating this complex substrate is the first hurdle. Experimentation with various wood types, mimicking the decomposition stage, and sterilization techniques are essential for establishing a suitable foundation for growth. One must consider the delicate balance of nutrients, mimicking the slow release provided by the natural decay process within the grove.

Temperature and humidity play critical roles. Shadowed groves, characterized by their stable microclimates, provide consistent temperature and humidity levels rarely replicated in standard laboratory conditions. Precision in maintaining these parameters is paramount. Fluctuations outside the narrow optimal range can inhibit growth, impacting both the overall health of the mycelium and its bioluminescent properties. Investing in environmental control chambers with precise monitoring and adjustment capabilities is highly recommended for serious cultivation attempts. Consider the subtle air currents within the grove that gently circulate air without excessive drying. Replicating this gentle airflow can be achieved with small, strategically placed fans within the growth chamber, further mimicking the natural environment.

Light exposure is a counterintuitive but critical factor. While these fungi emit light, they are not fond of direct sunlight. The dappled light of the forest canopy provides the ideal balance. Extended periods of darkness, punctuated by intermittent low-intensity light, are necessary to trigger and sustain bioluminescence. Simulating this natural light cycle within the lab requires specialized lighting equipment programmed to mimic the rhythmic shifts of day and night within the shadowed grove. Think beyond simple timers. Consider the gradual transitions of twilight and dawn, and aim to recreate this nuanced lighting profile for optimal results.

Sterility is essential. Contamination by other fungi or bacteria can quickly overwhelm the slow-growing bioluminescent mycelium. Maintaining a sterile environment throughout the cultivation process, from substrate preparation to inoculation and ongoing maintenance, is critical. This necessitates meticulous sterile techniques, including the use of a laminar flow hood, autoclaved equipment, and rigorous disinfection protocols. Consider the air quality within the growth chamber. HEPA filtration systems can further minimize the risk of airborne contaminants, protecting the delicate fungal cultures.

Patience is perhaps the most important ingredient. Unlike many commonly cultivated fungi, bioluminescent species often exhibit slower growth rates. It may take weeks, or even months, to observe significant mycelial development and bioluminescence. Regular monitoring, meticulous record-keeping, and adjustments to environmental parameters based on observations are key to success. Avoid the temptation to frequently disturb the cultures. Excessive handling can introduce contaminants and disrupt the delicate growth process. Trust the process and allow the fungi the time they need to establish themselves.

Documenting your trials is a crucial element of successful cultivation. Maintain a detailed log of every aspect of your process, from the specific substrate composition and environmental parameters to observations on growth rate, bioluminescence intensity, and any signs of contamination. This meticulous record-keeping allows for iterative refinement of your techniques, enabling you to optimize conditions for robust and consistent bioluminescence. Photographs and time-lapse imagery can provide valuable visual documentation of the growth process, capturing the subtle changes in mycelial development and light emission over time.

Beyond the basic requirements, enhancing bioluminescence is a further challenge. While simply achieving growth is a significant step, maximizing the intensity and duration of the glow requires deeper understanding. Research suggests that certain nutrients and environmental cues can influence bioluminescence. Experimentation with varying nutrient concentrations, the introduction of specific trace elements, and mimicking the natural stressors present in the shadowed grove can potentially unlock enhanced light production. Consider the role of symbiotic relationships within the grove. Exploring the potential for introducing beneficial bacteria or other microorganisms could further mimic the natural environment and potentially stimulate bioluminescence.

Cultivation trials are a journey of discovery, requiring patience, meticulousness, and a deep respect for the delicate balance of nature. By understanding the intricate relationship between these fascinating fungi and their environment, we can create the conditions for them to thrive and share their ethereal glow with the world. Remember that each species of bioluminescent fungi has unique requirements. What works for one may not work for another. Embrace the challenges and celebrate the small victories along the way, for the rewards of cultivating these luminous wonders are well worth the effort.

5.3 Preservation Efforts

The ethereal glow of bioluminescent fungi, captivating as it is, represents a fleeting beauty. These delicate organisms, often thriving in specific microclimates within shadowed groves, are vulnerable to environmental shifts and habitat degradation. Preserving these living lights requires a multifaceted approach, balancing scientific study with responsible conservation practices. A key component of this endeavor lies in meticulously documenting the precise environmental conditions in which these fungi flourish. This involves measuring factors such as temperature, humidity, soil composition, and surrounding flora. Understanding these intricate ecological relationships allows for the creation of targeted preservation strategies, ensuring the long-term survival of these luminous wonders. Think of it as crafting a bespoke environment, tailored to the specific needs of each species.

Beyond environmental parameters, genetic diversity plays a crucial role in the long-term resilience of bioluminescent fungi populations. Maintaining a robust gene pool helps these organisms adapt to changing environmental conditions and resist diseases. This necessitates careful management of existing populations and, where possible, the establishment of new populations in suitable habitats. Such initiatives might involve transferring spores or mycelial fragments to new locations, mimicking the natural dispersal processes that contribute to genetic exchange. Imagine facilitating the silent conversation between these fungal communities, fostering a network of resilience against future threats.

Cryopreservation, a technique involving freezing biological material at ultra-low temperatures, offers a promising avenue for preserving the genetic heritage of bioluminescent fungi. This method allows for the long-term storage of spores, mycelial cultures, and even tissue samples, creating a living library of genetic diversity. This library could prove invaluable in future restoration efforts or for scientific research, unlocking further secrets about these enigmatic organisms. It's akin to safeguarding a treasure trove of biological information, a repository of potential for generations to come.

Establishing protected areas within shadowed groves provides a critical safeguard against habitat loss, a primary threat to bioluminescent fungi. These designated zones act as sanctuaries, shielding the delicate ecosystems from human encroachment and destructive activities. Collaborating with local communities, landowners, and conservation organizations ensures the long-term success of these initiatives. Think of these protected areas as havens, where the soft glow of these fungi can continue to enchant for generations to come.

Ex-situ conservation, the practice of maintaining organisms outside their natural habitat, complements in-situ efforts, offering a safety net against unforeseen environmental catastrophes. Botanical gardens, research institutions, and specialized facilities can play a vital role in cultivating and preserving bioluminescent fungi under controlled conditions. This not only safeguards against extinction but also provides opportunities for research and public education, fostering appreciation for these extraordinary organisms. Envision these ex-situ collections as living embassies, representing the magic of the shadowed groves to a wider audience.

Citizen science initiatives, harnessing the power of public participation, offer a valuable tool for monitoring bioluminescent fungi populations and expanding our understanding of their distribution. By engaging amateur naturalists and enthusiasts, researchers can gather data from a broader geographical area and over longer time periods, uncovering patterns and trends that might otherwise go unnoticed. This collaborative approach empowers individuals to become active participants in conservation efforts, fostering a sense of shared responsibility for these luminous treasures. Imagine a network of eyes, attuned to the subtle glow of these fungi, contributing to a collective understanding of their hidden world.

Educating the public about the ecological importance of bioluminescent fungi and their fragile habitats is essential for fostering widespread support for conservation efforts. Through outreach programs, educational materials, and interactive exhibits, we can inspire a sense of wonder and appreciation for these remarkable organisms. This, in turn, translates into greater public engagement and advocacy for their protection. Think of it as illuminating the hidden connections between these fungi, their environment, and our own well-being.

Sustainable tourism, when carefully managed, can provide economic incentives for local communities to protect bioluminescent fungi and their habitats. By offering opportunities for guided tours and educational experiences, communities can benefit directly from the preservation of these natural wonders. This creates a positive feedback loop, where economic prosperity and conservation reinforce each other. Envision a future where the ethereal glow of these fungi becomes a source of both wonder and sustainable livelihoods.

International collaboration plays a crucial role in sharing knowledge and resources for the preservation of bioluminescent fungi across borders. By exchanging expertise and coordinating conservation efforts, researchers and conservationists can amplify their impact and address global challenges more effectively. Think of it as weaving a global network of support, dedicated to safeguarding these luminous organisms for the benefit of all.

Promoting responsible research practices ensures that scientific studies of bioluminescent fungi are conducted ethically and sustainably. This includes minimizing disturbance to natural habitats, obtaining necessary permits and approvals, and sharing research findings transparently. These ethical considerations uphold the integrity of scientific inquiry while contributing to the long-term preservation of these remarkable organisms. Consider it a commitment to responsible exploration, ensuring that the pursuit of knowledge does not come at the expense of the very organisms we seek to understand.

5.4 Ethical Considerations

The pursuit of artistic expression should never come at the expense of the natural world, particularly when dealing with organisms as delicate and elusive as bioluminescent fungi. Respect for their habitat is paramount. Venturing into shadowed groves requires a mindful approach, minimizing disturbance to the ecosystem. Footpaths should be adhered to, avoiding trampling of undergrowth or displacement of the delicate fungal colonies. Consider the cumulative impact of multiple visits to a single location and explore alternative sites to distribute the pressure on these sensitive environments. Responsible artistic endeavors prioritize preservation over exploitation.

Documenting the location and characteristics of bioluminescent fungi contributes valuable data to scientific understanding, but this must be done responsibly. Avoid revealing precise locations publicly to prevent over-collection or habitat destruction by less scrupulous individuals. Collaboration with researchers and conservation organizations can ensure that data collection aligns with established protocols and contributes to ongoing conservation efforts. The ephemeral nature of these luminous displays underscores the importance of non-destructive observation. Resist the temptation to collect specimens, as this can disrupt the delicate balance of the ecosystem and hinder the fungi's reproductive cycle. Photography and other non-invasive documentation methods offer ways to capture the beauty of these organisms without harming them.

Cultivating bioluminescent fungi presents a unique set of ethical dilemmas. While controlled environments may facilitate artistic expression and scientific study, they should be approached with caution. The provenance of cultures is crucial. Sourcing fungi from reputable suppliers who prioritize sustainable collection practices is essential. Introducing non-native species to an environment can have unforeseen consequences, potentially disrupting local ecosystems and outcompeting native fungi. Quarantine procedures and careful monitoring are necessary to prevent unintended releases. Furthermore, the genetic manipulation of bioluminescent fungi for artistic purposes raises ethical questions about altering natural processes and the potential for unintended ecological impacts.

The use of "stardust" and other artistic mediums requires careful consideration of its environmental impact. The sourcing and production of these materials should prioritize sustainability, minimizing the use of harmful chemicals or unsustainable resource extraction. Disposal of art materials must also be conducted responsibly, preventing contamination of natural environments. When working with bioluminescent fungi, ensure the medium itself does not interfere with the organism's natural processes or introduce toxins into its environment. Prioritize biodegradable and non-toxic materials whenever possible to minimize long-term environmental impact.

The commercialization of bioluminescent fungi and art derived from them raises further ethical considerations. Exploitation of these organisms for profit can incentivize unsustainable harvesting practices and threaten their long-term survival. Transparency in sourcing and production is crucial. Consumers should be informed about the origin of the fungi and the methods used in creating the art, empowering them to make ethical purchasing decisions. Supporting artists and businesses that prioritize sustainability and conservation is essential for ensuring the long-term preservation of these unique organisms. A portion of proceeds from art sales could be dedicated to funding research and conservation initiatives.

The very act of observing and interacting with bioluminescent fungi carries an ethical weight. These organisms are not merely objects of aesthetic appreciation but integral components of complex ecosystems. Approach them with humility and respect, recognizing the interconnectedness of all living things. Strive to educate and inspire others about the importance of conservation, fostering a sense of shared responsibility for protecting these luminous wonders and the shadowed groves they inhabit. Promote ethical practices within the artistic community and beyond. Engage in dialogue about the ethical implications of working with bioluminescent fungi. By fostering a culture of respect and responsibility, we can ensure that the pursuit of artistic expression contributes to the preservation of these extraordinary organisms for generations to come. The beauty of bioluminescent fungi should inspire not only artistic creation but also a deep commitment to protecting the natural world that sustains them. Ethical considerations are not merely constraints but opportunities to enhance the artistic process and deepen our connection with the natural world.

5.5 Future of the Fungi

The future of fungi, particularly bioluminescent species, hinges on a multifaceted approach that blends conservation, scientific exploration, and artistic inspiration. We stand at the precipice of a new era in our understanding of these organisms, an era where the ethereal glow of fungal bioluminescence can illuminate not only shadowed groves but also the path towards a deeper appreciation of the interconnectedness of life. Preserving existing habitats is paramount. This means identifying and protecting areas where bioluminescent fungi thrive, implementing sustainable forestry practices, and mitigating the impacts of climate change. These efforts require collaboration between scientists, conservationists, local communities, and policymakers, ensuring these unique ecosystems remain vibrant for generations to come.

Beyond habitat preservation, ex situ conservation plays a crucial role. Cultivating bioluminescent fungi in laboratory settings allows for controlled studies of their life cycle, physiology, and bioluminescent properties. This knowledge is essential for developing effective conservation strategies and for understanding the complex interplay of factors influencing their survival. Furthermore, ex situ cultivation can offer a sustainable alternative to harvesting wild populations for research or artistic purposes, reducing pressure on natural ecosystems. The development of robust cultivation techniques remains a challenge, requiring careful attention to substrate composition, humidity, temperature, and light exposure. Continued research into the specific needs of different species is crucial for optimizing cultivation protocols and ensuring the long-term viability of ex situ collections.

Unraveling the mysteries of fungal bioluminescence through scientific inquiry is equally vital. Investigating the biochemical pathways involved in light production, the genetic basis of bioluminescence, and the ecological role of this phenomenon will deepen our understanding of these remarkable organisms. Advanced molecular techniques, such as gene sequencing and transcriptomics, offer powerful tools for probing the intricate mechanisms underlying fungal bioluminescence. Understanding these processes could have far-reaching implications, potentially leading to the development of novel bioluminescent technologies for various applications, from medical imaging to sustainable lighting.

Furthermore, exploring the symbiotic relationships between bioluminescent fungi and other organisms within their ecosystems is critical. Understanding the interplay between fungi, plants, insects, and other organisms will provide insights into the ecological significance of bioluminescence and its role in shaping the delicate balance of these shadowed groves. Such research may also reveal the evolutionary pressures that have driven the development of bioluminescence in fungi, shedding light on the broader evolutionary history of this fascinating phenomenon.

The intersection of art and science presents a unique opportunity to foster a deeper appreciation for bioluminescent fungi. Artistic interpretations of these organisms, using mediums like stardust or other innovative approaches, can capture the public's imagination and inspire further scientific exploration. Art can serve as a powerful tool for communicating complex scientific concepts, bridging the gap between researchers and the wider community. By portraying the beauty and wonder of bioluminescent fungi through artistic expression, we can foster a sense of awe and inspire a collective desire to protect these extraordinary organisms and their delicate ecosystems. Collaborations between artists and scientists can lead to new forms of creative expression, informing artistic practice with scientific insights and enriching scientific understanding with artistic perspectives.

Ethical considerations must underpin all endeavors related to bioluminescent fungi. Responsible research practices, sustainable harvesting methods, and respect for the cultural significance of these organisms in local communities are crucial. Open access to research findings and data sharing are essential for fostering collaboration and advancing our understanding of these fungi. Developing guidelines for ethical research and artistic practices involving bioluminescent fungi will ensure that our interactions with these organisms are respectful and sustainable, benefiting both science and art while preserving the natural wonders of the shadowed groves. The future of fungi, especially those that illuminate the darkness with their ethereal glow, rests in our hands. By embracing a holistic approach that combines scientific inquiry, artistic expression, and ethical conservation, we can unlock the full potential of these remarkable organisms and ensure that their light continues to shine for generations to come. The path forward requires a shared commitment to preserving biodiversity, fostering scientific discovery, and inspiring a deeper appreciation for the natural world.

6 Chapter 6: Unveiling the Secrets

The ethereal glow of bioluminescent fungi has captivated scientists and artists alike for centuries. The precise biochemical mechanisms that allow these organisms to produce light, however, remain a source of ongoing investigation. Early research identified the presence of luciferin, a light-emitting molecule, and luciferase, the enzyme that catalyzes the oxidation of luciferin, as the key components in fungal bioluminescence. This process, similar to that found in fireflies and certain marine organisms, involves a complex interplay of enzymatic reactions that ultimately result in the emission of photons, the fundamental particles of light. Yet, unlike their insect counterparts, the specific types of luciferin and luciferase utilized by fungi, and the intricate regulatory pathways governing their activity, are still being unravelled. Understanding these processes is crucial not only for deciphering the mysteries of fungal bioluminescence but also for potential applications in bioimaging and biotechnology.

The ecological role of bioluminescence in fungi presents another intriguing puzzle. Several hypotheses have been proposed, ranging from attracting insects for spore dispersal to deterring fungivores. For instance, some researchers believe that the glowing mycelium of certain fungi attracts insects that then inadvertently carry fungal spores to new locations, aiding in the organism's propagation. Others posit that the light serves as a warning signal to potential predators, advertising the fungus's potentially toxic nature. Furthermore, the rhythmic pulsations of light observed in some species suggest a possible role in communication within the fungal colony itself, perhaps coordinating growth or metabolic processes. Observational studies in the natural habitats of these fungi, combined with laboratory experiments manipulating light exposure and insect interactions, are essential to further elucidate the ecological significance of this mesmerizing phenomenon. Beyond the biochemical pathways and ecological functions, the evolutionary history of bioluminescence in fungi presents a fascinating area of study. Phylogenetic analyses suggest that bioluminescence has evolved independently multiple times within the fungal kingdom, raising questions about the selective pressures that drove the emergence and persistence of this trait. Comparing the genetic makeup of luminous and non-luminous fungal species can provide insights into the evolutionary origins and diversification of the genes involved in light production. Tracing the evolutionary trajectory of bioluminescence can also shed light on the broader ecological and evolutionary dynamics of shadowed groves, the unique ecosystems where these glowing fungi thrive.

The spectral emissions of bioluminescent fungi, spanning from cool greens and blues to warmer yellows and oranges, also merit closer examination. While the dominant wavelengths emitted by different species are relatively well-documented, the factors influencing these variations remain largely unexplored. Environmental conditions, such as temperature and humidity, might play a role, as could the specific composition of the fungal substrate. Moreover, the intensity and duration of light emission can vary considerably, possibly reflecting internal biological rhythms or responses to external stimuli. Precise measurements of spectral emissions under controlled laboratory conditions, coupled with field observations across diverse habitats, are needed to fully understand the factors that shape the mesmerizing light displays of these extraordinary organisms.

The interaction of light with the surrounding environment within the shadowed grove adds another layer of complexity to the study of fungal bioluminescence. The dense canopy and leaf litter of these ecosystems create a unique light environment, filtering and scattering the bioluminescent glow. This interplay of light and shadow likely influences the visibility and perceived intensity of the fungal luminescence, impacting its ecological effectiveness. Furthermore, the reflectivity of surrounding surfaces, such as leaves and decaying wood, can contribute to the overall ambiance of the shadowed grove, creating an ethereal landscape where light and darkness intertwine. Detailed studies of light propagation and scattering within these environments, using advanced imaging techniques and light meters, are crucial for fully appreciating the visual impact of bioluminescent fungi.

Delving into the cellular structures responsible for light production in fungi offers further insights. Microscopic examination reveals specialized organelles within the fungal hyphae, the thread-like filaments that make up the fungal body, where bioluminescence occurs. These structures, often localized near the tips of the hyphae, contain the necessary machinery for producing and emitting light. Investigating the ultrastructure of these organelles using electron microscopy, and analyzing their protein composition using proteomic techniques, can provide a deeper understanding of the cellular mechanisms that underpin fungal bioluminescence.

Finally, the ongoing exploration of new species of bioluminescent fungi promises to unveil even more secrets. Recent expeditions to remote and unexplored regions have yielded discoveries of novel luminous fungi with unique characteristics, further expanding the known diversity of these fascinating organisms. Molecular phylogenetic analyses of these newly discovered species, combined with detailed characterization of their light-emitting properties, will undoubtedly enrich our understanding of the evolution and ecology of fungal bioluminescence. As we continue to explore the shadowed groves of our planet, we can anticipate uncovering even more wonders of the fungal world, illuminating the mysteries of light production in these remarkable organisms.

6.1 Scientific Inquiry

Understanding the bioluminescence of fungi requires a multifaceted approach, blending meticulous field observation with rigorous laboratory analysis. Begin your scientific inquiry by focusing on the ecological context. Observe the specific microclimates these fungi inhabit. Document factors like humidity, temperature, and the composition of the surrounding substrate, be it decaying wood, leaf litter, or other organic matter. Note the presence of other organisms, from insects to plants, as they might play a role in the fungi's lifecycle or light production. Detailed photographic documentation, using long-exposure techniques in low-light conditions, can provide valuable visual data, capturing the subtle nuances of the fungal glow and its interaction with the environment. This meticulous fieldwork forms the foundation for deeper investigations.

Transition from the field to the laboratory, where advanced techniques can illuminate the biochemical mechanisms behind bioluminescence. Employ high-performance liquid chromatography (HPLC) to isolate and identify the luciferins, the light-emitting molecules, and the luciferases, the enzymes that catalyze the light-producing reaction. Spectrometry helps determine the precise wavelengths of light emitted, providing insights into the spectral diversity of these fungi. Gene sequencing allows you to delve into the genetic basis of bioluminescence, identifying the genes responsible for luciferin and luciferase production, and potentially revealing evolutionary relationships between different bioluminescent species. Comparing the genetic makeup of closely related species, some luminescent and others not, can offer clues about the evolutionary origins and advantages of this fascinating trait.

Further your understanding by exploring the relationship between fungal bioluminescence and its ecological function. Design experiments to test hypotheses about the role of light. Does it attract insects for spore dispersal? Does it deter predators? Controlled experiments, using artificial light sources mimicking the fungal glow, can help determine the behavioral responses of insects and other organisms. Observe insect activity around bioluminescent fungi in their natural habitat, comparing it to activity around non-luminescent fungi in similar environments. Quantify the number and types of insects visiting each type of fungi, providing statistical evidence for or against a specific ecological role of bioluminescence. Consider the possibility of symbiotic relationships between the fungi and other organisms. Are there specific bacteria or insects that are consistently associated with bioluminescent fungi, potentially playing a role in their light production or dispersal? Microscopic analysis and culturing techniques can help identify and study these potential symbiotic partners.

Expand your inquiry beyond the individual organism to consider the broader ecological implications of bioluminescent fungi. How does their presence influence the overall biodiversity of the shadowed groves they inhabit? Do they contribute to nutrient cycling or play a unique role in the decomposition process? Long-term monitoring of fungal populations and their surrounding environment can provide valuable data on their ecological impact. Consider the potential effects of environmental changes, such as climate change or habitat destruction, on these delicate ecosystems and the bioluminescent fungi within them. By studying the specific environmental factors that influence fungal growth and luminescence, you can develop strategies for their conservation and preservation. This integrated approach, combining fieldwork, laboratory analysis, and ecological investigation, will provide a comprehensive understanding of the mesmerizing phenomenon of bioluminescent fungi. Remember, each observation, each experiment, adds another piece to the puzzle, illuminating the intricate interplay of life and light within the shadowed groves. Through rigorous scientific inquiry, you can contribute to unveiling the secrets of these luminous wonders and ensure their continued presence in our world. Don't forget to meticulously document all your findings, from field observations to laboratory data, creating a comprehensive record of your research journey. This meticulous documentation not only strengthens the validity of your conclusions but also provides a valuable resource for future researchers, fostering collaboration and advancing our collective understanding of these captivating organisms. Finally, communicate your findings to the wider scientific community through publications and presentations, contributing to the ongoing dialogue about bioluminescent fungi and their significance in the natural world. Investigate the evolutionary history of bioluminescence in fungi. Compare the genetic sequences of bioluminescent fungi to those of their non-luminescent relatives to understand the evolutionary pathways that led to the development of this trait. Examine the phylogenetic relationships between different bioluminescent fungal species to determine whether bioluminescence evolved independently multiple times or arose from a common ancestor. Exploring these evolutionary connections can offer insights into the environmental pressures and adaptations that drove the emergence of bioluminescence in the fungal kingdom. Consider the biogeographical distribution of bioluminescent fungi. Are they concentrated in specific regions, or are they scattered across the globe? Mapping their distribution can reveal patterns related to climate, habitat type, and evolutionary history. Analyzing the bioluminescent properties of fungal specimens collected from different geographical locations can further illuminate the diversity and adaptability of these organisms.

6.2 Mysteries of Light

The phenomenon of bioluminescence in fungi is more than just a captivating display; it's a complex interplay of biochemical reactions, ecological adaptations, and evolutionary history. We observe the light, but the true mystery lies in its origin, its purpose, and its variations. The luciferin-luciferase system, responsible for the light emission, is not uniform across all species. Subtle differences in the structure of these molecules lead to variations in the color of the emitted light, from cool greens and blues to warmer yellows and oranges. This biochemical diversity raises intriguing questions. Why do different species emit different colors? Does the color play a role in attracting insects for spore dispersal, or perhaps in deterring predators? These questions represent active areas of research, pushing the boundaries of our understanding.

Consider the ecological context of these glowing fungi. They inhabit shadowed groves, environments characterized by low light and high humidity. In this context, bioluminescence becomes a powerful tool. Some researchers hypothesize that the light attracts insects which then inadvertently spread the fungal spores, aiding in reproduction. Others propose that the light may serve as a warning signal to deter grazing animals, protecting the delicate fungal structures. The interplay of light and shadow, of organism and environment, creates a complex ecological web where bioluminescence plays a crucial, yet still partially understood role.

The temporal patterns of bioluminescence add another layer to the mystery. Some species glow continuously, while others exhibit pulsating or flashing patterns. What drives these rhythmic variations? Are they influenced by environmental factors such as temperature or humidity, or are they governed by internal biological clocks? The answers to these questions could provide insights into the fundamental mechanisms regulating fungal metabolism and behavior. Furthermore, understanding the specific environmental triggers that influence bioluminescence could aid in conservation efforts, enabling us to better protect these fragile ecosystems.

Beyond the biochemical and ecological aspects, the evolutionary history of bioluminescence in fungi remains an enigma. Did it evolve independently in different lineages, or does it share a common ancestor? Comparing the genetic makeup of bioluminescent and non-bioluminescent fungi could shed light on the evolutionary pathways that led to this remarkable adaptation. By tracing the origins of bioluminescence, we can gain a deeper appreciation for the interconnectedness of life on Earth and the incredible diversity of evolutionary strategies.

The study of fungal bioluminescence is an interdisciplinary endeavor, requiring expertise in biochemistry, ecology, genetics, and evolutionary biology. It is a field ripe with unanswered questions, beckoning researchers to delve deeper into the mysteries of these luminous organisms. New imaging techniques are allowing us to visualize the light emission at a microscopic level, revealing intricate patterns and interactions within the fungal tissues. Advanced genetic tools are providing insights into the genes responsible for bioluminescence and their regulation.

The potential applications of this research extend beyond the realm of basic science. Understanding the biochemical mechanisms of fungal bioluminescence could lead to the development of new bioluminescent markers for medical and environmental monitoring. Imagine using bioluminescent fungi to detect pollutants in soil or water, or to track the spread of disease in plants. These are just a few examples of the exciting possibilities that lie ahead. By continuing to explore the mysteries of light in the fungal kingdom, we not only expand our knowledge of the natural world, but also pave the way for innovative applications that could benefit society. As we uncover the secrets of these elusive organisms, we gain a greater appreciation for the intricate beauty and complexity of life itself. The pursuit of knowledge is a journey, and the mysteries of light beckon us forward, promising new discoveries and a deeper understanding of the world around us. The interplay of light and shadow in the shadowed groves becomes a metaphor for the intricate dance between the known and the unknown, inspiring further exploration and artistic interpretation.

7 Chapter 7: Art Meets Science

The convergence of artistic expression and scientific inquiry offers a unique lens through which to appreciate and understand the ethereal beauty of bioluminescent fungi. Where science seeks to dissect and categorize, art strives to interpret and connect. This intersection creates a powerful synergy, fostering a deeper appreciation for the natural world and inspiring new avenues for both artistic creation and scientific discovery. Consider the intricate network of a fungal mycelium, a hidden world revealed through scientific investigation. This intricate web, pulsing with life, becomes a source of artistic inspiration, translated onto the canvas through the medium of stardust, capturing its hidden luminescence.

The artistic process can serve as a powerful tool for scientific communication, transforming complex data into visually compelling narratives. Imagine a time-lapse photograph, captured over several nights, showcasing the rhythmic pulsations of a fungal colony. This visual representation not only captures the aesthetic beauty of the phenomenon but also provides valuable data on the fungal life cycle, its interaction with the environment, and potentially even its response to subtle environmental changes. This fusion of art and science allows researchers to share their findings with a broader audience, fostering public understanding and support for conservation efforts. The ethereal glow of the fungi, captured in a long-exposure photograph, can ignite the imagination and inspire further scientific inquiry.

Moreover, the act of artistic creation can spark new scientific questions. An artist, attempting to replicate the intricate patterns of fungal growth using stardust, might inadvertently stumble upon a new method for cultivating these elusive organisms. The artist's exploration of different substrates and application techniques could lead to discoveries about optimal growth conditions, contributing directly to scientific understanding. The delicate interplay of light and shadow in a stardust rendering might prompt a scientist to investigate the role of light in fungal bioluminescence, opening new pathways for research. The pursuit of artistic representation can lead to unexpected scientific breakthroughs, highlighting the fruitful exchange between these two seemingly disparate fields.

Furthermore, art has a unique capacity to engage emotions and inspire action. A stunning photograph of a bioluminescent fungal colony, delicately rendered in stardust, can evoke a sense of wonder and awe, fostering a deeper connection to the natural world. This emotional engagement can translate into tangible support for conservation initiatives, motivating individuals to protect these fragile ecosystems. A painting capturing the ethereal glow of a shadowed grove can raise awareness about the importance of preserving these unique habitats. The evocative power of art can be a potent force for environmental stewardship, bridging the gap between scientific understanding and public action.

Beyond its role in communication and conservation, art offers a unique perspective on the very nature of scientific inquiry. By embracing the subjective experience of observing and interacting with the natural world, art encourages scientists to look beyond the purely objective lens of data and analysis. The act of capturing the ephemeral beauty of bioluminescent fungi in an artistic medium requires a deep understanding of both the scientific principles governing their luminescence and the artistic techniques necessary to represent it. This interplay of objective observation and subjective interpretation enriches both the scientific and artistic processes, leading to a more holistic understanding of the natural world. The very act of choosing a specific shade of stardust to represent the fungal glow requires an artistic interpretation of scientific data, blurring the lines between the two disciplines.

Consider the artist meticulously applying stardust to a canvas, striving to capture the subtle gradations of light emitted by a cluster of fungi. This act of focused attention, of close observation and careful manipulation, mirrors the process of scientific investigation. The artist's careful selection of colors and textures to represent the fungal forms echoes the scientist's meticulous collection and analysis of data. This shared dedication to understanding and representing the natural world highlights the fundamental connection between art and science.

Finally, the integration of art and science in the study of bioluminescent fungi can lead to a more profound appreciation for the interconnectedness of life on Earth. The celestial echoes evoked by these luminous organisms remind us that we are part of a larger cosmic tapestry, a network of interconnected systems spanning from the microscopic to the macroscopic. By combining the rigor of scientific inquiry with the expressive power of art, we can gain a deeper understanding of our place within this vast and wondrous universe. The intricate patterns of fungal growth, mirrored in the swirling formations of distant nebulae, speak to the underlying unity of the cosmos. The artist's interpretation of these patterns, rendered in the shimmering medium of stardust, creates a tangible link between the terrestrial and the celestial, reminding us of our connection to the wider universe.

7.1 Bridging the Gap

The convergence of artistic expression and scientific inquiry offers a powerful synergy, particularly within the realm of bioluminescent fungi. We've explored the intricate biology of these organisms, the unique ecological niche they occupy, and the potential of stardust as a medium to capture their ephemeral beauty. Now, we'll delve into the fertile ground where art and science intersect, fostering a deeper understanding and appreciation for these luminous wonders. Artists, inspired by the ethereal glow of these fungi, can translate complex scientific concepts into visually compelling narratives, making them accessible to a wider audience. Imagine an installation where the rhythmic pulsations of fungal bioluminescence are translated into a mesmerizing light and sound display, effectively communicating the intricate signaling pathways within the fungal network. This translation not only captivates but educates, fostering a sense of wonder and respect for the natural world.

Consider the potential of scientific visualization techniques, transforming microscopic imagery of fungal hyphae and bioluminescent structures into large-scale art pieces. These magnified perspectives reveal hidden intricacies, allowing viewers to appreciate the complex architecture of these organisms. Perhaps a series of photographic prints, showcasing the diverse morphologies and spectral emissions of different fungal species, could illuminate the subtle variations in their light-producing mechanisms. Such an exhibit could spark curiosity and encourage further scientific investigation.

Furthermore, artistic interpretations of scientific data can generate new hypotheses and research directions. An artist's unique perspective may uncover patterns or connections overlooked by traditional scientific analysis. For instance, an artist exploring the spatial distribution of bioluminescent fungi within a shadowed grove might create a three-dimensional model highlighting the interplay of light and shadow. This artistic representation could then inspire scientists to investigate the influence of microclimates on fungal growth and bioluminescence. The cross-pollination of ideas between art and science enriches both disciplines, leading to unexpected discoveries.

The collaborative process itself fosters a deeper appreciation for the interconnectedness of knowledge. Scientists working alongside artists gain new insights into the aesthetic dimensions of their research, while artists gain a deeper understanding of the scientific principles underlying their creative endeavors. This mutual learning experience breaks down traditional disciplinary barriers, creating a more holistic approach to understanding the natural world. Imagine a workshop where mycologists and artists collaborate to create living sculptures using bioluminescent fungi. Such a project would not only produce stunning works of art but also contribute to scientific understanding of fungal growth and behavior in different environments.

This intersection of art and science also extends to conservation efforts. Art possesses a unique power to evoke emotional responses and inspire action. By portraying the beauty and fragility of bioluminescent fungi and their shadowed grove habitats, artists can raise awareness about the importance of preserving these ecosystems. A documentary film showcasing the elusive beauty of these fungi, interwoven with the scientific efforts to understand and protect them, could reach a global audience and inspire conservation initiatives. Perhaps a series of public murals depicting the interconnectedness of fungal networks with the surrounding environment could promote awareness and stewardship within local communities.

The artistic lens can also shed light on the ethical considerations surrounding the study and utilization of bioluminescent fungi. By exploring the potential impact of human intervention on these delicate ecosystems, artists can stimulate dialogue and encourage responsible practices. An interactive art installation, allowing viewers to virtually explore a shadowed grove and experience the impact of different conservation strategies, could serve as a powerful educational tool. The artistic narrative can personalize the scientific data, making the consequences of human actions more tangible and compelling.

Ultimately, the fusion of art and science creates a powerful platform for communication, education, and conservation. By bridging the gap between these two disciplines, we can unlock a deeper understanding of the natural world and inspire action to protect its wonders. The ethereal glow of bioluminescent fungi, captured and interpreted through the artistic lens, serves as a beacon, illuminating the intricate connections between the scientific and the aesthetic, the terrestrial and the celestial. It reminds us of the profound beauty and interconnectedness of life on Earth, urging us to safeguard these luminous treasures for generations to come. This shared future, where art and science walk hand-in-hand, promises a richer and more nuanced understanding of the natural world, fostering a deeper appreciation for the delicate balance that sustains life on our planet.

7.2 Inspiring Discovery

The convergence of art and science offers a unique lens through which we can appreciate and understand the ethereal beauty of bioluminescent fungi. Consider the meticulous field notes of a mycologist, painstakingly documenting the growth patterns, light emissions, and environmental conditions surrounding a newly discovered species. These observations, while scientifically valuable, can also inspire an artist to interpret the data visually, translating the complex scientific language into a captivating image that resonates with a wider audience. Imagine a series of photographs capturing the subtle shifts in luminescence throughout the night, transformed into an immersive installation that mimics the experience of walking through a shadowed grove. This artistic representation not only celebrates the beauty of the fungi but also communicates scientific findings in a powerfully accessible way.

Such collaborations can lead to unexpected breakthroughs. An artist's interpretation of fungal bioluminescence, for instance, might highlight patterns or variations in light emission that a scientist might overlook during purely quantitative analysis. This artistic perspective can prompt further scientific investigation, leading to new insights into the underlying mechanisms of bioluminescence, its ecological function, and its potential applications. Perhaps the artist's depiction of the intricate interplay of light and shadow within the grove reveals a previously unrecognized relationship between fungal luminescence and insect behavior, sparking a new avenue of ecological research. The artistic act of observation can thus become a catalyst for scientific discovery.

Moreover, this synergy extends beyond mere observation and interpretation. The development of "stardust," the artistic medium introduced earlier, exemplifies the potential of collaborative innovation. Imagine a specialized camera, designed in consultation with mycologists and artists, capable of capturing the full spectrum of fungal bioluminescence with unprecedented detail. This technology could not only provide scientists with valuable data but also empower artists to create breathtakingly realistic representations of these elusive organisms, bridging the gap between scientific understanding and artistic expression. Furthermore, stardust could be formulated to interact with fungal bioluminescence in novel ways, perhaps by enhancing or modulating its light emission, opening up entirely new avenues for artistic exploration and scientific experimentation.

The potential of this artistic medium transcends the purely aesthetic. Imagine using stardust to create interactive installations that respond to the presence of viewers, mimicking the natural rhythms of the fungal orchestras. Such installations could educate the public about the importance of fungal biodiversity and the fragility of shadowed grove ecosystems, fostering a sense of wonder and a desire to protect these remarkable organisms. Art, in this context, becomes a powerful tool for conservation, inspiring action by engaging audiences on an emotional and intellectual level.

The collaborative spirit between art and science can also manifest in the form of citizen science initiatives. Imagine a mobile app, designed by artists and scientists, that allows amateur mycologists to document and share their observations of bioluminescent fungi in their local environments. This data, collected and curated through the app, could contribute significantly to our understanding of the distribution, diversity, and ecology of these organisms, supplementing traditional scientific research with a vast network of citizen scientists. The app could also incorporate artistic elements, allowing users to create digital artwork inspired by their fungal encounters, further blurring the lines between scientific observation and artistic expression.

Consider the creation of a bioluminescent fungal sanctuary, a collaborative project between artists, scientists, and conservationists. This sanctuary could serve as a protected habitat for these elusive organisms, a living laboratory for scientific research, and an inspiring space for artistic creation. Imagine walking through a darkened grove, guided by the ethereal glow of bioluminescent fungi, surrounded by artwork that celebrates their beauty and illuminates their scientific significance. This immersive experience could foster a deep appreciation for the interconnectedness of art, science, and nature, inspiring a new generation of stewards for these luminous wonders. The synergy between art and science creates a positive feedback loop, where artistic insights drive scientific inquiry, leading to new discoveries that, in turn, inspire further artistic exploration. This continuous exchange of ideas and perspectives enriches both fields, fostering a deeper understanding and appreciation of the natural world. By embracing this collaborative spirit, we can unlock the full potential of bioluminescent fungi as both a source of scientific wonder and an inspiration for artistic creation, ensuring their preservation for generations to come. It's through this intertwined exploration that we truly begin to grasp the intricate harmonies within these shadowed groves, connecting us to the celestial sphere reflected in their gentle glow. This continuous cycle of discovery and creation, fueled by the interplay of art and science, offers a powerful pathway towards a deeper understanding and appreciation of the natural world, fostering a sense of wonder and a commitment to preserving these luminous treasures for generations to come. The future of our relationship with bioluminescent fungi, and indeed with the natural world as a whole, hinges on our ability to embrace this synergistic approach, celebrating the interconnectedness of scientific inquiry and artistic expression.

7.3 A Shared Future

The convergence of art and science presents a unique opportunity to foster a deeper appreciation for the natural world, particularly for organisms as elusive and captivating as bioluminescent fungi. By intertwining scientific inquiry with artistic expression, we can illuminate the mysteries of these organisms while simultaneously inspiring conservation efforts. Imagine a darkened grove, the air thick with humidity, where scattered clusters of bioluminescent fungi paint the forest floor with an ethereal glow. This otherworldly scene, captured through the lens of both the scientist and the artist, becomes a powerful narrative for the interconnectedness of life and the importance of preserving these fragile ecosystems. Scientists, with their meticulous observations and data collection, provide the foundational knowledge of fungal biology, their light-emitting mechanisms, and the delicate ecological balance within the shadowed groves. Artists, in turn, translate this scientific understanding into evocative visual and auditory experiences, capturing the ephemeral beauty and sparking a sense of wonder in a wider audience.

This collaborative approach allows us to transcend the purely academic realm and engage with the public on an emotional level. A photograph capturing the spectral emissions of a rare species, a sculpture mimicking the intricate structure of a fungal colony, or a musical composition echoing the rhythmic pulsations of fungal light—these artistic interpretations can ignite a passion for conservation where scientific papers might not. They bridge the gap between scientific knowledge and public awareness, transforming complex data into accessible and emotionally resonant narratives. Think of the impact of a time-lapse video showcasing the growth and luminescence of a fungal colony, set to a haunting melody inspired by the sounds of the grove. Such a piece can transport viewers into this hidden world, fostering a sense of connection and responsibility.

The shared future we envision is one where scientific research informs artistic creation, and where art, in turn, fuels further scientific exploration and conservation initiatives. Artists can participate in fieldwork, accompanying scientists into the shadowed groves to experience firsthand the magic of these bioluminescent organisms. This immersive experience can inspire new forms of artistic expression, capturing the essence of the fungal world in ways that traditional methods might miss. Imagine an artist creating a bioluminescent installation using cultivated fungi, a living artwork that changes and evolves over time, reflecting the dynamic nature of the organism itself. This form of "living art" could become a powerful tool for environmental education and advocacy.

Furthermore, collaborations between artists and scientists can lead to the development of innovative conservation strategies. Artists can help design educational programs, create compelling visual materials for outreach campaigns, and develop interactive exhibits that engage the public with the wonders of bioluminescent fungi. By weaving together scientific knowledge and artistic vision, we can create a powerful tapestry that inspires action and fosters a shared commitment to protecting these unique ecosystems. Consider the possibility of an augmented reality experience that overlays digital recreations of bioluminescent fungi onto real-world environments, allowing users to explore the hidden beauty of these organisms in their own backyards. Such a project could ignite a sense of curiosity and wonder, encouraging people to become active participants in conservation efforts.

This symbiotic relationship between art and science has the potential to unlock new avenues for understanding and appreciating the natural world. By working together, scientists and artists can create a more holistic and engaging narrative about the importance of biodiversity conservation, particularly for these often-overlooked luminescent wonders. The resulting synergy can lead to a deeper public understanding of the intricate relationships within ecosystems and the urgent need to protect them. Imagine a future where bioluminescent fungi are no longer elusive secrets hidden within shadowed groves, but rather celebrated icons of biodiversity, inspiring awe and wonder in people around the world. This shared future, built on the foundations of scientific discovery and artistic expression, is a future where the harmonies of nature and human creativity converge to create a more vibrant and sustainable world. The stardust, as a medium, can encapsulate this synergy, reflecting the delicate interplay of light, life, and art in the shadowed groves. It can serve as a visual testament to the shared future we strive for – a future where the ethereal glow of bioluminescent fungi continues to inspire both scientific inquiry and artistic expression for generations to come. Through this collaborative approach, we not only unlock the secrets of these fascinating organisms but also ensure their continued presence in the celestial tapestry of life on Earth.

7.4 Conservation Through Art

Art, in its multifaceted forms, possesses a unique power – the ability to transcend mere aesthetics and inspire action. Specifically, it can serve as a potent tool for conservation, bridging the gap between scientific understanding and public awareness. By capturing the ethereal beauty of bioluminescent fungi through artistic mediums like photography, sculpture, and even music inspired by their light patterns, we can foster a deeper appreciation for these organisms and their fragile ecosystems. This appreciation, in turn, can translate into tangible support for conservation efforts, driving funding, research, and protective legislation. Think of the iconic photographs of endangered species – images that have galvanized public support and spurred critical conservation initiatives. Bioluminescent fungi, with their otherworldly glow, hold a similar potential to captivate and inspire.

The very act of creating art centered around these fungi necessitates a deep engagement with their environment. Artists venturing into shadowed groves to capture their luminescence must understand the delicate balance of these ecosystems. They must learn about the specific conditions required for fungal growth, the interconnectedness of the fungal networks with other organisms, and the threats these habitats face. This intimate knowledge, gained through artistic exploration, becomes embedded within the artwork itself, subtly educating viewers and fostering a sense of stewardship. A photograph capturing the delicate interplay of light and shadow within a grove, a sculpture mimicking the intricate structure of a fungal fruiting body, or a musical composition echoing the rhythmic pulses of fungal bioluminescence can all convey a deeper understanding of these organisms and their importance.

Moreover, art can reach audiences far beyond the scientific community. Scientific papers, while crucial for research and understanding, often remain confined within academic circles. Art, on the other hand, has a universal language, capable of transcending cultural and educational barriers. A stunning image of a glowing mushroom can evoke wonder and curiosity in anyone, regardless of their scientific background. This accessibility makes art a powerful tool for raising public awareness about the importance of conserving these often-overlooked organisms. Imagine a public art installation showcasing the beauty of bioluminescent fungi, accompanied by information about their ecological significance and the threats they face. Such initiatives can spark public dialogue and inspire action.

Furthermore, the artistic process can offer new perspectives and insights into the scientific realm. Artists often approach subjects with a different lens than scientists, focusing on aesthetic qualities, emotional responses, and symbolic interpretations. This unique perspective can lead to novel observations and questions, ultimately enriching scientific understanding. For instance, an artist's fascination with the varying colors and intensities of fungal bioluminescence might inspire scientists to investigate the underlying biochemical mechanisms responsible for these variations. This cross-pollination of ideas can lead to breakthroughs in both artistic and scientific fields.

Integrating artistic representations of bioluminescent fungi into educational programs can also have a profound impact on conservation efforts. By incorporating engaging visuals and interactive art projects, educators can ignite students' curiosity and foster a deeper appreciation for the natural world. This early exposure can cultivate a lifelong commitment to environmental stewardship, inspiring future generations to protect these unique ecosystems. Picture a classroom where students create their own "stardust" inspired by fungal bioluminescence, learning about the scientific principles behind the phenomenon while simultaneously expressing their creativity.

Finally, art can be a powerful catalyst for fundraising and resource allocation for conservation initiatives. Art exhibitions, auctions, and collaborative projects can generate revenue to support research, habitat restoration, and community engagement programs. By showcasing the beauty and ecological significance of bioluminescent fungi through art, we can attract financial support from individuals and organizations committed to environmental protection. Imagine a limited-edition print of a stunning photograph of a bioluminescent mushroom, with proceeds going towards the preservation of its habitat.

In essence, art serves as a bridge between the scientific understanding of bioluminescent fungi and the public's appreciation for their beauty and importance. By engaging our emotions and sparking our curiosity, art empowers us to become active participants in their conservation. It transforms scientific knowledge into a tangible, accessible, and inspiring call to action, ensuring the survival of these luminous wonders for generations to come. The intricate interplay of light and shadow within a shadowed grove, transformed into a captivating photograph, can speak volumes about the fragility and wonder of these unique ecosystems.

7.5 Celestial Echoes

The ethereal glow of bioluminescent fungi has captivated artists and scientists alike, inspiring a quest to bridge the gap between objective observation and subjective interpretation. Think of the shadowed grove as a natural cathedral, its vaulted canopy filtering moonlight onto the forest floor where these fungal constellations emerge. The interplay of light and shadow creates an otherworldly ambiance, a stage set for nature's own light show. This aesthetic experience transcends mere observation; it's a visceral connection to the mysteries of the natural world, prompting a deeper exploration of the intricate relationship between art and science. We seek not only to understand the biological mechanisms behind bioluminescence but also to capture its essence, its ephemeral beauty, and translate it into a tangible form. This is where the concept of "celestial echoes" comes into play.

Consider the artist as a translator, a conduit between the unseen world of fungal networks and the human eye. The artist doesn't merely replicate the visual spectacle of bioluminescence; they interpret it, imbuing it with personal meaning and emotional resonance. Stardust, as an artistic medium, becomes the language of this translation, a means of capturing and amplifying the faint whispers of the grove. Imagine capturing the subtle variations in light intensity, the pulsating rhythms of fungal colonies, and the intricate interplay of colors. Stardust allows us to freeze these fleeting moments, transforming them into lasting works of art that resonate with the celestial sphere they evoke. It's not simply about documenting the phenomenon; it's about creating a dialogue, a conversation between the observer and the observed.

Through the lens of art, we gain a new perspective on the scientific intricacies of bioluminescent fungi. Artistic representations can highlight patterns and relationships that might otherwise go unnoticed, inspiring new avenues of scientific inquiry. For instance, an artist's interpretation of the spatial distribution of fungal colonies could spark research into the underlying ecological factors influencing their growth patterns. Conversely, scientific understanding can enrich the artist's palette, providing a deeper appreciation for the complexities of the subject matter. Understanding the chemical reactions that produce bioluminescence, for instance, can inform the artist's choice of materials and techniques, allowing them to create works that are both aesthetically compelling and scientifically informed. This reciprocal exchange fosters a deeper understanding of the natural world, enriching both artistic expression and scientific discovery.

The concept of "celestial echoes" extends beyond mere representation. It embodies the interconnectedness of life on Earth and the vastness of the cosmos. The intricate mycelial networks that connect fungal colonies beneath the forest floor mirror the intricate web of relationships that bind together all living organisms. These networks, invisible to the naked eye, play a crucial role in the health and resilience of forest ecosystems. Just as the stars in the night sky appear to communicate across vast distances, the fungal networks facilitate communication and resource sharing within the hidden world beneath our feet. By exploring these hidden connections, we gain a deeper appreciation for the intricate tapestry of life on Earth. Art becomes a tool for unveiling these hidden connections, for making the invisible visible.

Ultimately, the convergence of art and science through the medium of stardust offers a powerful tool for conservation. By capturing the beauty and wonder of bioluminescent fungi, we can raise awareness of their ecological importance and the threats they face from habitat loss and environmental degradation. Art can inspire a sense of stewardship, motivating individuals and communities to protect these delicate ecosystems. Imagine a series of art installations that immerse viewers in the ethereal glow of bioluminescent fungi, evoking a visceral connection to these organisms and their shadowed grove habitats. Such experiences can transcend the purely intellectual and tap into the emotional core of our relationship with the natural world, fostering a deeper sense of responsibility for its preservation. The "celestial echoes" of these remarkable organisms remind us that we are all interconnected, part of a larger cosmic web. By embracing the synergistic relationship between art and science, we can unlock new pathways to understanding, appreciating, and protecting the wonders of the natural world.

www.ingramcontent.com/pod-product-compliance
Ingram Content Group UK Ltd.
Pitfield, Milton Keynes, MK11 3LW, UK
UKHW020158290926
15471UKWH00005B/940

9 798348 384692